D1213328

Survival Skills for the Fire Chief

Survival Skills for the Fire Chief

ROBERT S. FLEMING, ED.D.

PennWell®

Fire Engineering®

Disclaimer: The recommendations, advice, descriptions, and the methods in this book are presented solely for educational purposes. The author and publisher assume no liability whatsoever for any loss or damage that results from the use of any of the material in this book. Use of the material in this book is solely at the risk of the user.

Copyright© 2011 by
PennWell Corporation
1421 South Sheridan Road
Tulsa, Oklahoma 74112-6600 USA

800.752.9764
+1.918.831.9421
sales@pennwell.com
www.Fire EngineeringBooks.com
www.pennwellbooks.com
www.pennwell.com

Marketing Coordinator: Jane Green
National Account Executive: Cindy Huse

Director: Mary McGee
Managing Editor: Marla Patterson
Production Manager: Sheila Brock
Production Editor: Tony Quinn
Book Designer: Susan E. Ormston
Book Layout: Lori Duncan
Cover Designer: Karla Pfeiffer

Library of Congress Cataloging-in-Publication Data

Fleming, Robert S.
 Survival skills for the fire chief / Robert S. Fleming.
   p. cm.
 Includes index.
 ISBN 978-1-59370-256-4
 1. Fire departments--Administration--Vocational guidance. I. Title.
 TH9158.F594 2011
 363.37068--dc22
                    2010047157

Printed in the United States of America

1 2 3 4 5   15 14 13 12 11

To my loving family, whose support and encouragement have allowed me to live my dream of being a fire officer and instructor, as well as to share my experience through writing this book.

Contents

Preface

Over the past 30 years, I have had the unique opportunity to engage in two distinct but related careers—one in fire and emergency services administration, and one as an academician on the business faculty at a number of colleges and universities. Most individuals consider themselves fortunate if they are able to experience one meaningful career. I have had the distinct privilege of having two. An interesting note is that my endeavors in both arenas have been complementary and synergistic and have enabled me to engage in extensive action research on the management and leadership of contemporary fire departments.

This book has been written to meet the needs of a specific audience: those members of the fire service who aspire to become a fire chief and those currently serving in this mission-critical position within their fire department. It is organized around 26 distinct but interrelated topics that I have identified as mission-critical areas in determining the success of the contemporary fire chief and that of his or her fire department. These topical areas were selected based on personal and professional experience and validated through ongoing research on the management of contemporary fire departments.

My intent in writing this book was that it would serve as a "value-added" source of information for the aspirant or incumbent fire chief. It is designed and organized to serve as a resource as one prepares for the position of fire chief, providing information and insights that will be beneficial to members of the fire service taking promotional examinations or participating in assessment centers. It will likewise serve as a valuable and informative job aid and reference for the incumbent fire chief, as well as other fire department officers. It will also serve as a primary or supplemental textbook in related fire officer training and academic courses.

The book is designed to address the management mistakes that fire chiefs often make that compromise their effectiveness and ability to manage and lead their fire departments. Each of the 26 lessons contained in this book address a specific management survival skill and begins with an overview of the importance of that particular skill to the professional success and survival of the fire chief, as well as that of his or her fire department. This introductory material is followed by a concise presentation of what you need to know about each skill. Each lesson concludes with the delineation of the things that you will want to do, as well as those that you will not want to do, with respect to each management survival skill. Relevant job aids are provided to assist the reader in processing each lesson and applying it to their situation and that of their fire department.

I view the opportunity to share, through the writing of this book, the knowledge, skills, experience, and wisdom that I have gained over the past four decades in the fire service as a privilege and have endeavored to do so in a manner that will make the reader's time spent reading this book a value-added and meaningful experience. It utilizes a format that integrates theory and practice in a manner that enables both those preparing to become fire chief and those currently serving in this crucial position to benefit from reading this book. Lastly, I trust that your fire service career will be as interesting and rewarding as mine has been. And I hope that this book will assist in your preparation for the job and will contribute to your professional success and that of the fire departments that have entrusted you with the responsibility of managing and leading.

About the Author

Dr. Robert S. Fleming, a Professor of Management at Rowan University, is a highly respected author, instructor, conference speaker, researcher, and consultant within the fire and emergency services. He had been actively involved in fire and emergency services administration for more than 39 years and has served in numerous operational and administrative positions, including that of fire chief. He has served as a chief officer in three fire departments, currently holding the rank of battalion chief, training and professional development in the Goshen Fire Company of West Chester, Pennsylvania.

In addition to a Doctorate of Education in Higher Education Administration, Dr. Fleming has five earned master's degrees, including a Master of Business Administration from Temple University and a Master of Government Administration from the Fels Center of Government of the University of Pennsylvania. The primary focus of his research, teaching, and consulting has been on the enhancement of organizational effectiveness, with an emphasis on local, county, state, regional, and national fire and emergency services organizations.

His professional activities have included serving on the National Fire Academy (NFA) Board of Visitors for 13 years, including 6 years as vice chairman and 6 years as chairman. Dr. Fleming is a veteran member of the NFPA 1021 Technical Committee on Fire Officer Professional Qualifications. His current professional activities include serving as chairman of the Commonwealth of Pennsylvania Fire Service Certification Advisory Committee and the Chester County Local Emergency Planning Committee. He is a member of the Certified Fire Protection Specialist Board and previously served as its chairman.

Dr. Fleming is a certified fire instructor in Pennsylvania and New Jersey. He has contributed to all of the major fire service publications, has developed and presented numerous management and officer development programs for the Fire and Emergency Television Network (FETN), and has delivered sessions on fire and emergency services administration and fire officer development at the major fire service conferences. Through his teaching, conference presentations, and writing, he has trained and mentored numerous members of the fire service as they prepared for and assumed the position of fire chief.

His numerous professional certifications and designations include chief fire officer (CFO), certified fire protection specialist (CFPS), and executive fire officer (EFO). He is also the author of *Effective Fire and Emergency Services Administration*, published by PennWell/Fire Engineering Publishing.

Lesson 1: Making the Transition to Fire Chief

Role in Survival and Success

Assuming the position of fire chief within a contemporary fire department represents a pinnacle in your fire service career. Your career pilgrimage likely began as a firefighter, with initial advancement to the rank of line officer, then to chief officer, and finally to the position of fire chief. As a firefighter, your responsibilities involved performing technical work under the direction of a line officer.

Fire Department Career Progression

- Firefighter
- Line officer
- Chief officer
- Fire chief

As a line officer, your responsibilities on the incident scene were to ensure the safety of the firefighting crews you supervised in their performance of tactical assignments (figure 1–1). Your off-scene responsibilities as a line officer may have included assignments such as station training, preplanning, public education, apparatus and equipment maintenance, and station maintenance (figure 1–2). Whereas your technical skills remained invaluable as a line officer, you probably quickly realized that rather than following your natural instincts to jump in and perform a task yourself, your new role required that you develop the ability to prepare and rely on those under your supervision to do the work.

Figure 1–1. Line officer supervising firefighting crew. (Courtesy of Bob Sullivan)

Figure 1–2. Line officer conducting in-station training

In advancing to the chief officer ranks, your roles and responsibilities once again changed. While as a line officer, you served in a supervisory capacity; as a chief officer you were expected to manage and lead the fire department and its members (figures 1–3 and 1–4). As a chief officer, depending on your rank and responsibilities, you were a member of either middle or top management.

Figure 1–3. Fire chief commanding incident. (Courtesy of Tony Barbato)

Figure 1–4. Fire chief making presentation. (Courtesy of Dan Jones)

Each of the previously referenced illustrations of fire service career advancement represents a transition. As you advance from firefighter to line officer, to chief officer, and finally to fire chief, it is imperative that you recognize that you are making a transition. The key to a successful transition is to recognize that you have assumed a new set of roles and responsibilities and consequently undertaken significant professional and personal change. External candidates who accept employment within an organization usually recognize that they are making a transition from their previous position and organization to a new position and organization. They also understand that this requires making the necessary adjustments to succeed in their new position and organization. It is often the case, however, that those advancing within an organization, including to the position of fire chief, fail to recognize that they, too, are making a transition and thus fail to take all of the necessary actions that will contribute to their success in this new position.

Representative Chief Officer Ranks

- Battalion chief
- Assistant chief
- Deputy chief
- Fire chief

Your success, and often survival, as a fire chief will in large part be determined by your ability to make the transition to your new position effectively. A successful transition begins with having a realistic preview of the roles and responsibilities of your new position as fire chief. Failure to recognize that you are making the ultimate transition within your fire department and act accordingly will likely compromise your professional success as fire chief, as well as that of the fire department that you have been entrusted with the opportunity to manage and lead.

What You Need to Know

Some of the knowledge, skills, and wisdom that you gained from your previous transitions to and through the fire officer ranks will certainly be valuable in contributing to your success and survival as a fire chief; however, the challenges associated with this final transition can be significantly greater in many instances. An essential aspect of this transition requires you to have a full understanding and appreciation of the department or municipal process by which individuals are appointed or elected to the position of fire chief within your fire department. This process and its accompanying procedures vary by fire department and municipality, as well as among volunteer, combination, and career fire departments.

Whereas your preparation as a firefighter focused on mastery of the technical tasks required of a position incumbent and included initial and refresher training in firefighting and related disciplines, this preparation was supplemented with supervisory training prior to or upon assuming a line officer position. Likewise, given that the primary role of a chief officer is to effectively, efficiently, and safely manage and lead his or her fire department, preparation in these areas is central to preparing the chief officer and his or her fire department for professional and organizational success.

In addition to the skills that you needed while serving as a line officer—including planning, organizing, directing, controlling, communication, decision making, and problem solving—making a successful transition to the chief officer ranks requires that a position incumbent develop the necessary knowledge, skills, and attitudes to effectively lead, motivate, and empower others. You will quickly learn the importance of mutual respect, wherein you develop a respect for others while ideally having them also develop a respect for you.

You need to develop a skill set that enables you to provide the essential visionary leadership to successfully manage and lead your fire department during a time of new and unprecedented challenges, along with the challenges that fire departments and their chiefs have faced for years. The importance of providing the necessary professional development opportunities, whether in the form of training, education, or certification, to the members of your fire department and in particular to those who currently serve as fire officers, as well as those who aspire to do so, must be emphasized.

Similar to the lesson that you learned as a line officer regarding your role to supervise rather than perform tactical tasks, you must now realize the need to delegate appropriate responsibilities to your fire officers, particularly to those serving as chief officers. To do so is an essential aspect of the fire officer's professional development, as well as a management survival skill. Delegation enables you to spend your time on the essential roles and responsibilities of fire chief that include providing visionary leadership in strategic planning, creating a positive and empowering organizational climate, budgeting and financial management, and establishing external relations with fire department stakeholders.

Your professional success and that of your fire department require that you provide visionary leadership that focuses on the future, rather than dwelling on the past and present. Your roles and responsibilities, in all likelihood, involve a substantial increase in your activities outside your fire department as you interact with and relate to its stakeholders. The significance of external roles is likely to increase as you assume the position of fire chief, as are the challenges associated with balancing internal and external roles.

As a fire chief, you have a responsibility with respect to stewardship in terms of handing over an improved fire department to the individual who eventually succeeds you. It is essential that you manage and lead your fire department in a proactive, rather than a reactive, manner. It is also imperative that you avail yourself of professional development opportunities, and resist the tendency to feel that you do not need additional training and education or that you do not have the time to participate in these professional development opportunities. Whereas the practical experience that you are gaining each and every day during your service as fire chief certainly has the potential of enhancing your ability to succeed, not taking advantage of appropriate professional development opportunities can potentially compromise your professional success as fire chief, as well as that of your fire department.

Job Aid 1 that appears at the end of this lesson has been provided to assist you in formulating a realistic understanding of the roles and responsibilities of the fire chief within your department and enable you to develop an understanding of the knowledge and skills that you need to succeed in this crucial position.

Things to DO:

- Recognize that you are making a major transition in assuming the position of fire chief.

- Understand the internal and external roles of the fire chief, and strike an appropriate balance between these two important sets of roles that will continually compete for your time and attention.

- Understand the organization-specific processes for the appointment and retention of the fire chief within your jurisdiction.

- Understand the stakeholder expectations of those appointing or electing the fire chief, fire department members, and the community.

- Practice integrity, and demonstrate ethical decision making when enacting all responsibilities.

- Engage in appropriate professional development opportunities, including training, education, and certification, both in preparing for the position of fire chief and during your tenure as fire chief.

- Provide appropriate professional development opportunities to members of your fire department, particularly those currently serving in or aspiring to officer positions, and encourage their participation in these opportunities.

- Serve as a mentor to fire department members as they make the transition to and through the officer ranks, particularly those who ultimately aspire to the position of fire chief.

- Surround yourself with a highly qualified cadre of officers.

Things NOT to Do:

- Fail to recognize that you are making a transition of greater significance and consequence than your prior advancement to and through the fire officer ranks.

- Fail to recognize that your roles and responsibilities as fire chief will be significantly different than those of the fire officer positions that you previously held.

- Engage in decision making or other activities that compromise your integrity; engage in unethical behavior, or violate established laws and regulations.

- Fail to take advantage of professional development opportunities by rationalizing that you do not need professional development or are too busy to engage in these opportunities.

Making a Successful Transition to Fire Chief (Job Aid 1)

1. Describe the characteristics of the roles and responsibilities of fire officer positions you have held prior to becoming fire chief.

2. Describe the characteristics of the roles and responsibilities of fire chief within your fire department.

3. Identify the challenges or obstacles to making a successful transition to fire chief within your fire department.

4. Identify the things you could do to enhance the success of your transition to the position of fire chief.

Lesson 2: Challenges You Will Face as Fire Chief

Role in Survival and Success

An individual assuming a senior management position in any contemporary organization quickly learns that accompanying the defined roles and responsibilities of the position are a growing number of challenges that the organization and the individual at its helm must understand, anticipate, and be prepared to address properly. Some of these challenges have likely been present in the organization's operating environment for a period of time; however, some are recent, and the astute manager should expect additional challenges in the future. An essential starting point in understanding the present and future challenges that you will face as fire chief and that will confront your fire department is to fully understand, recognize, and acknowledge that the contemporary fire department is an organization and thus requires the same professional approach to management and leadership as does any other contemporary organization. An organization is an organized structure of roles and responsibilities wherein individuals, working in groups or teams, perform the work of the organization in furtherance of its mission or purpose.

> **Organization**
>
> An organized structure of roles and responsibilities wherein individuals, working in groups or teams, perform the work of the organization in furtherance of its mission or purpose

As fire chief, you are expected to provide the necessary guidance to skillfully combine the department's tangible and intangible resources with its capabilities in a manner that contributes to the distinctive competencies required to fulfill the fire department's mission and meet and, where possible, exceed the expectations of its stakeholders. Tangible resources are physical resources, such as land, plant, apparatus, equipment, and capital. Intangible resources include the department's reputation, goodwill, and image in terms of public perception of the organization. Capabilities are the potential of an organization to successfully utilize its tangible and intangible resources.

Elements of Organizational Success

- Resources
- Tangible resources
- Intangible resources
- Capabilities
- Distinctive competencies

Although your fire department will certainly need to address some common challenges that contemporary organizations face, such as economic and financial realities as well as recruiting and retaining highly qualified personnel, there will also be many unique challenges that you and your fire department will face based on the nature of your enterprise and the mission-critical services that it provides to the community. An illustrative example of this would be the fact that the population of the United States is aging. In a fairly short time, we as a nation, including fire departments and particularly those that currently provide emergency medical services or enter this field of service delivery in the future, will experience a significant increase in the demand for services that are logically associated with an aging population. The information available through the U.S. Census Bureau and other sources will provide essential guidance as you prepare and position your fire department to address such demographic challenges (figure 2–1).

Figure 2–1. Census reports. (Data from U.S. Census Bureau)

As fire chief, your ability to provide the required visionary leadership to position your organization to effectively, efficiently, and safely meet the emergency service needs of the stakeholders that you serve is instrumental in determining both your success and the success, and perhaps survival, of your fire department. The lessons in this book have been developed in recognition of the challenges that you and your department will likely face in the present as well as in the future. They are designed to provide you with the necessary knowledge, skills, and insights that will contribute to your professional success as a fire chief.

What You Need to Know

It is imperative that you fully understand the roles and responsibilities that accompany the position of fire chief. You need to recognize that you are a senior manager entrusted with managing and leading your fire department. Your roles and responsibilities will be multifaceted and encompass many areas. Your success will, in large part, be influenced by your understanding of the stakeholders of your fire department and their expectations, as well as your actions to position the organization to meet and, where possible, exceed reasonable stakeholder expectations. The unique mission of the contemporary fire department and the resulting services that it provides present a number of challenges to the fire department and those charged with its successful operation. These challenges relate to the nature of the services provided by the fire department in that they typically involve the intangible delivery of labor intensive services, which are time critical, with immediate consumption. The challenges of fire department service delivery are further complicated by the reality of unscheduled service delivery and varying transaction volumes.

Challenges in Delivering Fire Department Services

- Immediate service consumption
- Intangible delivery
- Labor intensive
- Time-critical services
- Transaction volume
- Unscheduled service delivery

Through a proactive approach, you can develop a full and comprehensive understanding of the current status of your fire department and the environment in which it operates. Such a proactive approach to environmental scanning and subsequent strategic planning will enable you to position your organization to build on its strengths and minimize its weaknesses while pursuing strategic environmental opportunities and avoiding environmental threats.

Organizational and Environmental Considerations

- Organizational strengths
- Organizational weaknesses
- Environmental opportunities
- Environmental threats

An analysis of your fire department will reveal a number of key components that fall on a continuum, ranging from major strengths to major weaknesses. These items, which should be considered in ascertaining and addressing challenges you either currently face or will face in the future, include: apparatus, equipment, facilities, financial situation, human resources, internal climate and morale, leadership, and response capacity and readiness, as well as other organizational factors. Potential opportunities or threats likewise fall on a continuum and include: changing laws, regulations, and standards; changing social and cultural norms; demographic changes; fundraising potential; municipal support; new initiatives; new technologies; political developments; and redistricting, regionalization, and consolidation.

An essential component of the success of any organization and the achievement of its mission is its ability to recruit, motivate, empower, and retain personnel with the knowledge, skills, and attitudes necessary to perform the organization's work. As a fire chief you will find that you will be involved in more human resource management matters than you perhaps bargained for when taking the position. It is important that each position within your fire department has an accompanying job description that defines the duties and responsibilities of that position and an accompanying job specification that details the necessary qualifications to successfully enact that position. Human resource management activities, which may at times prove challenging, will relate to the recruitment, retention, motivation, empowerment, discipline, and evaluation of fire department personnel.

As a fire chief, you will utilize management skills in planning, organizing, directing, and controlling. In so doing, you will also be engaging in decision making and communication. There will be times when you will be required to utilize your management talents in problem solving or complaint or conflict management. Your leadership role as fire chief will require you to provide the proactive and visionary leadership necessary to get the stakeholders of your fire department, including its members, to reach a consensus on a desired future direction for your fire department and an appropriate set of strategies to pursue that future state. Essentially, you will often be serving as a change agent within your fire department and, at times, in its external environment.

As fire chief, you will be expected to enact appropriate stewardship in all of your actions and decisions and should be held accountable for this responsibility. Three essential areas of resource management are financial resources, human resources, and physical resources (figures 2–2 and 2–3). Each of these areas will present its own set of challenges during your tenure as fire chief.

Resource Management Challenges

- Managing financial resources
- Managing human resources
- Managing physical resources

Figure 2–2. Fire department personnel. (Courtesy of West Chester Fire Department)

Figure 2–3. Fire department station and apparatus. (Courtesy of Kevin Carney)

You must strive, despite the many anticipated or unforeseen challenges that you may encounter as fire chief, to enhance your organization and its ability to more fully meet and, where possible, exceed the realistic expectations of its stakeholders. In so doing, you must ensure that both you, as fire chief, and the organization that you manage and lead are fully compliant with all relevant laws and regulations, as well as demonstrating ethical behavior and integrity in all activities, dealings, and actions. A failure to do so will likely result in devastating consequences to your fire department in terms of its reputation, image, and public support. Such inappropriate actions usually also have corresponding devastating personal and professional consequences for the fire chief. The fact that a fire department is an extension of government, which provides essential public safety services, enhances the visibility of the fire department and often places it and its leaders in a "fishbowl" in terms of media coverage and visibility. As an individual who desires to achieve professional success as a fire chief, while also enhancing your fire department's success, you will absolutely want to avoid such problems through enacting your roles and responsibilities in a professional and ethical manner. An essential part of ensuring your fire department's success will be to fully utilize available avenues to market and promote your organization.

Job Aid 2 at the end of this lesson has been provided to assist you in formulating a realistic understanding of the present and future challenges facing your fire department and the corresponding challenges that you will be required to confront and address as fire chief.

Things to DO:

- Recognize that your fire department is an organization and that, like other successful contemporary organizations, it requires professional management and leadership.
- Provide visionary leadership that defines a desired strategic direction and associated organizational strategies.
- Develop a thorough and comprehensive understanding of the current challenges that your fire department is facing, along with an informed understanding of likely future challenges.
- Position your fire department to successfully address current challenges and prepare for future challenges.
- Utilize environmental scanning to identify organizational strengths and weaknesses, as well as environmental opportunities and threats.
- Develop strategic plans for the future that are mission-driven and build on organizational strengths, minimize organizational weaknesses, pursue environmental opportunities, and avoid environmental threats.
- Function as a change agent, motivating others to recognize and respond to the need for change within your fire department.
- Ensure that necessary strategies are in place to contribute to effective recruitment, motivation, empowerment, and retention of fire department personnel.
- Enact all roles and responsibilities as fire chief in a manner that demonstrates legal and regulatory compliance, integrity, and ethical behavior.
- Utilize all available opportunities to market and promote your fire department and the services it provides to the community it serves.

Things **NOT** to Do:

- Fail to recognize that you are a senior manager entrusted with the roles and responsibilities of successfully managing and leading a contemporary organization.

- Manage in a reactive, rather than proactive, manner that results in not properly positioning your fire department to address present and future challenges.

Present and Future Challenges for the Fire Chief and Fire Department (Job Aid 2)

1. Identify the present challenges that your fire department is facing.

2. Identify the future challenges that your fire department is likely to face.

3. Identify the present challenges that you face as fire chief.

4. Identify the future challenges that you are likely to face as fire chief.

Lesson 3: Roles of the Fire Chief

Role in Survival and Success

The key to success in any position begins with having a thorough understanding of its roles and responsibilities and the accompanying expectations of organizational stakeholders. Your roles and responsibilities as fire chief will differ from those of the previous positions that you held within the fire department. Although some of your roles and responsibilities will be enacted on the incident scene, such as commanding an emergency incident or interacting with the media at that incident, a fairly significant component of your work as fire chief will be enacted off the incident scene as you prepare your fire department to effectively, efficiently, and safely serve the community and meet and, where possible, exceed the reasonable expectations of your department's stakeholders.

As fire chief, many of your roles will be enacted within your fire department, whereas others will be performed outside the fire department. It is important that you fully understand both of these sets of roles and determine and enact an appropriate balance between your internal and external roles. The specific roles and responsibilities of the fire chief within your department are determined by a number of factors, including your job description, the needs of the community that your department serves and protects, the expectations of stakeholders, and the degree to which you delegate to other fire officers. You will quickly realize the importance of effective delegation as a management survival skill, and you must always appreciate that you are ultimately responsible for your fire department's operation and performance. You must also understand that there are some responsibilities that the astute fire chief will retain, rather than delegate to subordinates.

The roles and responsibilities that have been defined for the position of fire chief are the measures by which your performance and success are evaluated. These roles and responsibilities are thus instrumental in the accountability process, by which you are judged in terms of how successful you are in enacting your responsibilities and utilizing your accompanying authority as you manage and lead your fire department in an effective, efficient, and safe manner.

What You Need to Know

As you advance through the management ranks of any organization, including a fire department, your specific roles and responsibilities will change. As a line officer involved in supervising a crew operating on the incident scene or during a public education program, your responsibilities were rather limited in scope. Your successful enactment of your roles and responsibilities as a line officer required that you possessed and utilized appropriate technical and human skills. Now, as fire chief, it is your charge to manage and lead your fire department to success in the present, while preparing and positioning it to succeed in the future. Although effective human skills will still be extremely important as you enact the roles and responsibilities of fire chief, conceptual skills, wherein you can see the big picture and focus on the overall organization and its environment, will become crucial.

Your roles and responsibilities as fire chief, as well as your time, will be apportioned between inside and outside roles. Inside roles include those tasks that you perform within the fire department and on the incident scene. Outside roles include those tasks that you accomplish outside the fire department, within and sometimes beyond your community, as a representative of your fire department or the larger fire service.

Inside versus Outside Roles

- Inside roles
- Outside roles

The understanding of management roles was enhanced through Henry Mintzberg's identification of ten managerial roles. A manager may perform as few as one or as many as ten of these prescribed roles, based on the roles and responsibilities defined by an organization for a given position. Mintzberg categorized the ten managerial roles into three role sets: interpersonal roles, informational roles, and decisional roles.

Managerial Role Sets (Mintzberg)

- Interpersonal roles
- Informational roles
- Decisional roles

Interpersonal roles involve working or interacting with others and include the roles of figurehead, liaison, and leader. As a figurehead, you will perform ceremonial and symbolic duties, such as attending a function as a departmental representative (figure 3-1). When you interact with other organizations and agencies on the behalf of your fire department, you are functioning in the liaison role. Your activities, as fire chief, to motivate and empower members within your department fall under the role of leader.

Interpersonal Roles (Mintzberg)

- Figurehead
- Leader
- Liaison

Figure 3–1. Fire chief representing department at community event. (Courtesy of Susan Wentz)

Managerial roles that involve the management of information, called informational roles, include monitor, disseminator, and spokesperson. When you gather information about proposed regulations or standard revisions, you are enacting the role of monitor (figure 3–2). The disseminator role includes the subsequent sharing of that information within your fire department. When you communicate information to individuals, groups, or organizations outside your fire department, you are serving as a spokesperson.

Informational Roles (Mintzberg)

- Monitor
- Disseminator
- Spokesperson

Figure 3–2. Fire chief reviewing new standard. (Courtesy of Frank Hand)

The final set of managerial roles proposed by Mintzberg, decisional roles, includes entrepreneur, disturbance handler, resource allocator, and negotiator. When you, as fire chief, seek opportunities to position your fire department to better fulfill its mission and meet, and hopefully exceed, stakeholder expectations, you are enacting the role of entrepreneur. Your handling of disciplinary matters within the fire department or resolving citizen complaints would be an example of the disturbance handler role. The decisions you make regarding the setting of priorities in allocating the scarce resources of your fire department would represent the role of resource allocator. There will be times when you will be called upon to engage in formal and/or informal negotiations on behalf of your fire department, thus enacting the role of negotiator.

Decisional Roles (Mintzberg)

- Entrepreneur
- Disturbance handler
- Resource allocator
- Negotiator

As fire chief, you will be expected to enact your roles and responsibilities both on and off the incident scene. Although at times your responsibilities off the incident scene, whether involving attending or conducting meetings, preparing and managing fire department budgets, developing and implementing strategic plans, or other administrative tasks, may seem unglamorous and unimportant, such essential activities serve as the foundation for the effective, efficient, and safe response to emergency incidents. This essential work includes ensuring that your fire department is prepared to respond to emergency incidents, regardless of nature, type, or size, in an effective, efficient, and safe manner that fully meets and, where possible, exceeds the reasonable expectations of the stakeholders within your community. It is your responsibility to ensure that your fire department has the necessary response capacity and readiness in terms of personnel, apparatus, and equipment. Your main priority in all that you do as fire chief must be to ensure the life safety of your department's personnel as well as the community. An integral part of your role both on and off the incident scene should, therefore, be to prevent personnel injuries and fatalities, as well as address community risks.

Job Aid 3 that appears at the end of this lesson has been provided to assist you in developing a thorough understanding of the roles and responsibilities of the fire chief within your fire department.

Things to DO:

- Recognize that your roles and responsibilities as fire chief are different than those associated with previous positions that you held within the fire department.

- Recognize that as fire chief you are responsible for managing and leading your fire department in a manner that contributes to its operational effectiveness, efficiency, and safety.

- Develop a thorough understanding of your specific roles and responsibilities as fire chief, as well as the expectations of fire department stakeholders for your fire department and for you as fire chief.

- Recognize your responsibility to prepare and position your department to succeed in both the present and the future.

- Recognize that life safety must be your priority, and act accordingly as you enact your roles and responsibilities as fire chief.

- Utilize effective delegation as appropriate in the interest of more fully utilizing organizational resources, including the talents of subordinates while affording them professional development opportunities and freeing up your time to focus on essential responsibilities.

- Fully understand and enact your interpersonal roles as figurehead, leader, and liaison.

- Properly enact your informational roles, including monitor, disseminator, and spokesperson.

- Understand and, as appropriate, enact your decisional roles as an entrepreneur, disturbance handler, resource allocator, and negotiator.

Things NOT to Do:

- Fail to fully understand the roles and responsibilities that your fire department and its stakeholders expect of the fire chief.

- Focus on the small things in the present rather than the larger conceptual issues necessary to ensure your fire department's future success and survival.

- Fail to realize the importance of your roles and responsibilities off the incident scene in determining operational effectiveness, efficiency, and safety on the incident scene.

- Fail to fully understand the appropriate inside and outside roles of the fire chief and to set your priorities and allocate your time in a manner that reflects an appropriate balance between these conflicting role demands.

- Delegate tasks or responsibilities to subordinates that should be properly retained and enacted by you as fire chief, based on the nature of the task or stakeholder expectations with respect to the position of fire chief.

Identifying Your Roles and Responsibilities as Fire Chief (Job Aid 3)

1. Identify the primary roles of the fire chief within your fire department.

2. Identify the major responsibilities of the fire chief within your fire department.

3. Identify the roles of fire chief that you are most prepared for and confident in enacting.

4. Identify the responsibilities of fire chief that you are most prepared for and confident in enacting.

Lesson 4: Your Role as a Manager

Role in Survival and Success

An essential component of your roles and responsibilities as fire chief involves the management of your organization. Management involves working with and through others to accomplish mutual objectives. Management is considered both an art and a science. As a science, it involves the application of management theories, decision-making models, and tools; as an art, it involves having the wisdom and judgment to manage in an appropriate manner in a given situation. The skillful use of effective delegation is an example of the art of management, which, when used appropriately, will enable you to more effectively and efficiently manage and lead your fire department.

Desirable Organizational Outcomes

- Effectiveness
- Efficiency
- Safety

Your success as a manager will in large part be determined by and evaluated in terms of your ability to effectively and efficiently manage the human, financial, and physical resources available to your fire department. Your management approach and results will be key measures that organizational stakeholders will consider when evaluating your success as fire chief and whether or not you should continue to serve in this mission-critical capacity. Recognizing the importance of management in your portfolio of roles and responsibilities as fire chief and enacting this role in a professional manner that demonstrates a commitment to the fire department and its mission of service to the community, as well as a commitment to managing the organization with which you have been entrusted in a manner that demonstrates integrity and stewardship, will stand you in good stead as the senior manager within your fire department.

What You Need to Know

Management is a process of working with and through others to accomplish mutual objectives on behalf of the organization and its members. As a process, it is composed of a sequential set of activities, or management functions, that are performed over time. The management process also represents a cycle in that after the last activity, the process frequently reverts back to the first.

Management

Process of working with and through others to accomplish mutual objectives

As a fire chief, you will perform the management functions of planning, organizing, directing, and controlling towards the end of fulfilling your department's mission and achieving its goals in an effective, efficient, and safe manner. In so doing, you will use the integral skills of decision making and communication as you perform each of the four management functions. These interrelated management functions serve as the building blocks of your professional success, as well as the success of your fire department.

Management Functions

- Planning
- Organizing
- Directing
- Controlling

Management Support Activities

- Communication
- Decision making

Communication

Exchanging ideas or thoughts in the interest of achieving shared understanding

Decision Making

Process of making informed decisions based on available information

Planning is the process of establishing goals and an appropriate course of action or strategies to achieve these goals. Planning is considered the primary management function, in that it precedes the other management functions and lays the foundation for effective and efficient management. The other management functions sequentially follow planning and build on its foundation. As fire chief, you will quickly recognize that it is always important to have a plan before you proceed with an initiative, whether managing a major incident or designing a new fire apparatus or fire station.

Planning

Setting a direction for an organization or project

Planning

The primary management function

As you advanced through the officer ranks, you likely had some exposure to, and hopefully involvement in, departmental planning initiatives and their implementation. Although planning takes place at all levels of a fire department, the nature, scope, and time horizon of planning activities increase as one climbs the organizational ladder. As fire chief, your planning activities should focus on positioning your department for present and future success and survival through enhancing its ability to meet and, where possible, exceed reasonable stakeholder expectations. It is also your role to ensure that the planning initiatives throughout your fire department are coordinated and in support of the defined organizational mission.

Although you must certainly be concerned about the day-to-day operations of your fire department, your primary focus in planning must be to plan for the future through the facilitation of an organized and inclusive strategic planning process. The products of this planning effort, including a mission statement that provides essential strategic direction, goals, and objectives that support the organizational mission, and the strategies necessary to achieve the goals and objectives, are of paramount importance to the successful management of any contemporary organization, including your fire department.

A well-conceived and realistic plan that is based on a thorough understanding and incorporation of the comprehensive insights gained through the use of environmental scanning will enable your fire department, under your direction and leadership, to develop a multi-year strategic plan designed to build on organizational strengths, while minimizing organizational weaknesses, and pursue environmental opportunities while avoiding environmental threats (figure 4–1). Your success in facilitating a strategic planning process and the resulting strategic plan will provide essential guidance in terms of a desired destination, strategic goals, and route of travel identified as strategies as you seek to advance your organization and position it for continued success. In addition to the essential direction that the strategic plan will provide your administration during its tenure, such a plan will serve to demonstrate now and in the future the commitment and stewardship of your administration.

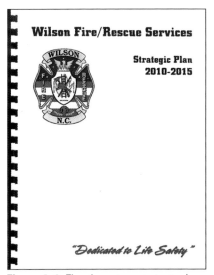

Figure 4–1. Fire department strategic plan. (Courtesy of Wilson Fire/Rescue Services)

Organizing, the second management function, logically follows the primary management function of planning. Organizing involves the acquisition and deployment of organizational resources to accomplish the goals, and ultimately the mission, of your fire department. Just as you organize and assign fire apparatus and personnel when managing an incident as the incident commander, there will be numerous decisions that you will make off the incident scene with respect to organizing.

Organizing

Allocating resources in support of a plan

The plans developed during your planning activities will provide instrumental guidance with respect to your resource needs and how to best allocate the scarce resources available to your fire department. The reality of scarce resources has become a more pronounced and unfortunately ever-present management challenge in most contemporary fire departments. Therefore, making the case to secure needed resources, including personnel, apparatus, equipment, and facilities, may prove to be one of the greatest challenges that you will face as fire chief (figure 4–2).

Figure 4–2. Fire department resources. (Courtesy of Bob Sullivan)

The directing function logically follows organizing and precedes controlling. It is through directing, or as it is sometimes called implementing, that the work of the organization is accomplished. The successful enactment of this crucial management function requires the implementation of the strategies developed during planning through the use of the organization's resources as defined, structured, and assigned through the organizing function. Your success in this area, as well as that of your fire department, will be grounded in the recognition that organizational members play an instrumental role in the implementation of any organizational strategy or initiative. The importance of motivating and empowering organizational personnel should be obvious in contributing to the successful execution of the management function of directing.

Directing

Process of implementing plans

Controlling involves the comparison of actual results or outcomes with planned results. Controlling utilizes measurable desired outcomes delineated during planning processes, such as goals and objectives, to evaluate organizational performance. The control process that you utilize should include the following sequential steps: establish goals, establish performance standards, measure actual performance, compare performance to standards, take corrective action as necessary, and provide feedback.

Controlling

Comparing results to goals and objectives and taking corrective action

Steps in the Control Process

1. Establish goals.
2. Establish performance standards.
3. Measure actual performance.
4. Compare performance to standards.
5. Take corrective action as necessary.
6. Provide feedback.

As a fire chief, you can employ three types of control within your fire department: feedforward control, concurrent control, and feedback control. Ensuring that all personnel are properly trained based on your department's scope of operations is an example of the use of feedforward control. Concurrent control takes place during the delivery of services, such as when you engage in continuous size-up of incident conditions and progress throughout the management of an emergency incident. The use of a post-incident analysis after a major incident is an example of feedback control. Review of departmental performance reports is another example of the use of feedback control (figure 4–3).

Types of Control

- Feedforward control
- Concurrent control
- Feedback control

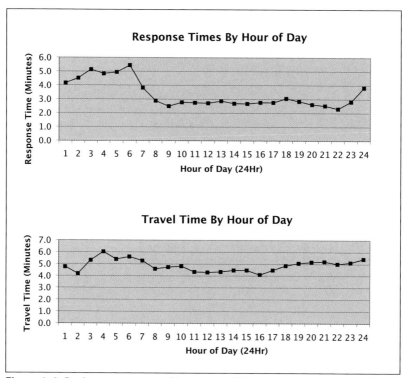

Figure 4–3. Performance report. (Courtesy of Rick Grothmann)

Job Aid 4 that appears at the end of this lesson has been provided to assist you in identifying the responsibilities and activities associated with enacting your management roles as the fire chief of your fire department.

Things to DO:

- Recognize the importance of effectively and efficiently managing your fire department's human, financial, and physical resources.

- Utilize effective delegation to enhance your effectiveness, efficiency, and success in managing your fire department.

- Recognize that as fire chief you have a mission-critical role in managing your fire department.

- Recognize that the integrated management functions of planning, organizing, directing, and controlling, supported by communication and decision making, are the building blocks of your professional success, as well as that of your fire department.

- Always remember that planning is the primary management function and should precede all other functions.

- Utilize strategic planning to establish essential strategic direction for your fire department, to prepare and position it for present and future success, and to enhance its ability to meet and, where possible, exceed realistic stakeholder expectations.

- Facilitate a strategic planning process that provides for the inclusion of all relevant fire department stakeholders, including fire department members.

- Utilize environmental scanning to reveal the necessary understanding and insights that will serve as the basis for a realistic strategic plan.

- Ensure that the strategic plan developed by your fire department builds on your fire department's strengths while minimizing its weaknesses and takes advantage of environmental opportunities while avoiding environmental threats.

- Recognize that fire department members play an instrumental role in the implementation of any strategies and initiatives and the resulting importance of motivating and empowering fire department personnel.

Things NOT to Do:

- Attempt to do everything yourself rather than delegate appropriate tasks and responsibilities to subordinates.

- Fail to demonstrate stewardship and integrity in all resource management decisions.

- Fail to understand that successful management is a process of working with and through others to accomplish mutual objectives.

- Engage in resource acquisition and deployment strategies that do not correspond with your fire department's strategic plan.

- Fail to recognize the reality of scarce resources and to demonstrate stewardship and integrity in making and justifying resource requests.

- Fail to implement and utilize organizational control measures and systems that incorporate feedforward, concurrent, and feedback controls.

Implementing Your Role as a Manager (Job Aid 4)

1. Identify your management responsibilities as fire chief with respect to planning.

2. Identify your management responsibilities as fire chief with respect to organizing.

3. Identify your management responsibilities as fire chief with respect to directing.

4. Identify your management responsibilities as fire chief with respect to controlling.

Lesson 5: Your Role as a Leader

Role in Survival and Success

Leadership is the process of influencing the behavior of others. Although your success as a fire chief will require you to demonstrate management skills, the true measure of your success will often come from your ability to not only manage, but provide the visionary leadership required of the contemporary fire chief and expected by fire department stakeholders. Your leadership will play an instrumental role in determining the success of your fire department and correspondingly your professional success as fire chief. As the fire chief of a contemporary fire department, you are called and expected to be a leader.

> **Leadership**
>
> Process of influencing the behavior of others

Through your service as fire chief, you will develop a thorough appreciation of the importance of your leadership in preparing and positioning your fire department for present and future success and to meet and, where possible, exceed the realistic expectations of your department's stakeholders. You will find that the members of your fire department represent a primary stakeholder group that will prove integral to the success and survival of your department. Whereas you manage fire department personnel using the power and authority granted to you upon your appointment as fire chief, you will discover that it is necessary to use different sources of power or influence to lead others, including fire department members. Your professional success as fire chief will be based on the recognition that fire department personnel follow your orders and directives because you have management authority over them (figure 5–1), whereas should they follow your direction as a leader, they are doing so because they want to follow you. Your success and survival as a fire chief, as well as that of your fire department, requires that you become an effective manager and leader.

Figure 5–1. Fire chief issuing orders on incident scene. (Courtesy of Bob Sullivan)

What You Need to Know

Management and leadership are both essential to the present and future success of your fire department. Your abilities and skills in these two crucial areas will determine your success as a fire chief, as well as your corresponding career potential. Throughout your career as a fire chief, you will face challenging situations that require you to be an astute manager and leader and select the appropriate management/leadership style for each situation. Your professional goals should, therefore, include becoming an effective manager and leader.

A starting point in becoming an effective manager and leader is to understand the similarities and differences between management and leadership. Management is the process of working with and through others to accomplish mutual objectives. Managers perform the management functions of planning, organizing, directing, and controlling in order to accomplish the organization's mission in an effective, efficient, and safe manner. Managers use the integral skills of communication and decision making as they enact these management functions.

The major similarity between management and leadership is that both approaches result in organizational members performing the work of the organization and working towards accomplishment of its goals. The primary difference between management and leadership is what motivates organizational members to follow the direction they receive. Individuals follow managers because they have to, based upon the power or authority that a superior holds; they follow leaders because they want to, based on other means of influence possessed by the leader. Leadership thus results from followership and willing compliance.

Management versus Leadership

- Fire department members comply with the orders of managers because they must.
- Fire department members and others follow the direction of leaders because they desire to do so.

Leadership

Leadership is based on followership and willing compliance.

As a fire chief, you want to avail yourself to a number of sources of power associated with management and leadership. Power is defined as the ability to influence the behavior of others. The five potential sources of power which you can utilize to influence the behavior of others are: legitimate power, reward power, coercive power, referent power, and expert power.

Sources of Power

- Legitimate power
- Reward power
- Coercive power
- Referent power
- Expert power

Legitimate power is granted by the organization as the authority associated with the position that you hold. Reward power is based on the ability to grant others things they desire, including assignments and rewards. Coercive power is the ability to discipline or punish others. Referent power, also referred to as personal power or identification power, is based upon a desire of others to identify or work with an individual, often as a result of respect or admiration. Expert power is based on an individual's knowledge, skills, and expertise.

Managers gain their power from three sources: legitimate power, reward power, and coercive power. Leaders have power attributed to them by followers through two sources: referent power and expert power. Upon appointment or election to the position of fire chief, you became a manager. Whether you also become a leader is up to you. Your success as a fire chief and that of your fire department will be greatly enhanced through your becoming a "manager/leader."

Sources of Power—Managers

- Legitimate power
- Reward power
- Coercive power

Sources of Power—Leaders

- Referent power
- Expert power

As a manager/leader, you have available to you all five sources of power. As a manager, you are able to use your legitimate, reward, and coercive powers to accomplish the responsibilities that you and those who report to you have been assigned. As a leader, you use referent power and expert power to motivate and empower these same individuals to want to follow your direction. In summary, as a manager/leader, you essentially have the best of both worlds at your disposal as you manage and lead your fire department (figure 5–2).

Sources of Power—Manager/Leader

- Legitimate power
- Reward power
- Coercive power
- Referent power
- Expert power

Figure 5–2. The fire chief as manager/leader. (Courtesy of Wilson Fire/Rescue Services)

As you manage and lead people, it is extremely important that you do so in a legal and ethical manner and that your decisions and treatment of others reflect honesty and integrity. Through managing in this manner, you will attract followers and thus become capable of leading and becoming a manager/leader. You must always strive to ensure that you use the powers that you have been granted by your organization or its members properly and prudently.

The adoption of an appropriate leadership style or approach will be central to your success as a leader. Although there are those individuals who practice autocratic leadership, where the leader seeks no or limited input from subordinates and simply tells them what to do, your success will be enhanced though the use of democratic leadership, also called participative management, wherein the leader seeks the input, involvement, and participation of organizational members. This approach has demonstrated its effectiveness in creating followership and reducing resistance to change on the part of organizational members.

Engage in transformational leadership by seeking to involve others in planning and change processes in the interest of transforming and revitalizing your fire department to more fully meet and, where possible, exceed realistic stakeholder expectations while achieving the organization's mission in an effective, efficient, and safe manner. As a successful fire chief, you will find it necessary to determine and utilize appropriate management and leadership styles in many different, and often challenging, situations. Adopt a mindset in dealing with these situations wherein you recognize that there is not one universal leadership style or approach that will be effective in all situations and that an appropriate leadership style must be ascertained and utilized in each situation.

Job Aid 5 that appears at the end of this lesson will assist you in your pilgrimage to become both a manager and leader within your fire department and the community that it serves.

Things to DO:

- Recognize and fully enact your role as a leader within your fire department and the community that it serves.
- Recognize the important role that leadership plays in determining your professional success and that of your fire department.
- Understand and utilize the sources of power available to managers.
- Understand and utilize the sources of power available to leaders.
- Recognize the differences in reasons why individuals follow the direction of managers and that of leaders.
- Purpose to become a manager/leader and act accordingly in your management/leadership approach.
- Manage and lead in a legal and ethical manner that builds trust, respect, and confidence.
- Purpose to become a transformational leader who prepares and positions your fire department to effectively, efficiently, and safely achieve its mission while meeting and, where possible, exceeding realistic stakeholder expectations.
- Recognize the importance of adjusting your leadership style based on situational factors.

Things NOT to Do:

- Focus only on being a manager and relinquish your leadership role.
- Fail to understand the similarities and differences between management and leadership.
- Use the powers that you have been granted by the fire department or its members in an inappropriate manner.
- Manage and lead in an autocratic manner that does not encourage or value the input, involvement, or participation of other members of your fire department.

Becoming a Manager/Leader
(Job Aid 5)

1. Describe how, as fire chief, you can use legitimate, reward, and coercive powers in managing your fire department.

2. Describe how, as fire chief, you can use referent and expert powers in leading your fire department.

3. Identify the challenges you face in becoming a manager/leader.

4. Describe the strategies you plan to use in becoming a manager/leader.

Lesson 6: Preparing Yourself for Success

Role in Survival and Success

Your professional success as a fire chief, as well as that of the fire department that you have been entrusted with managing and leading, will in large part be determined by your preparation before assuming the position of fire chief and the continuous professional development that you engage in throughout your tenure as fire chief. The importance of continuous professional development to maintain and enhance your knowledge and skills cannot be overemphasized.

Whereas your responsibilities as fire chief will often be demanding and keep you extremely busy, it is essential that you continue to recognize the importance of maintaining the currency and relevancy of your knowledge and skills, as well as expanding your knowledge and skill set, particularly in areas that will enhance your ability to successfully manage and lead your fire department in a contemporary world that continually presents new challenges. The reality is that many fire chiefs recognize the need for, and take time to participate in, numerous professional development opportunities before assuming the position of fire chief, but for various reasons they fail to continue their commitment to professional development after attaining the position of fire chief. The fallacy of so doing must be understood because it compromises your full potential to successfully continue to manage and lead your fire department, and prepare and position it for the present and future. In contrast, adopting a mindset that values and seeks professional development opportunities will prove beneficial to your success as fire chief and to that of your fire department.

What You Need to Know

As a fire chief, you will face many traditional, as well as new and unprecedented, challenges as you seek to manage and lead your fire department in a manner that contributes to achieving its mission of providing service to the community in an effective, efficient, and safe manner that meets and, where possible, exceeds realistic stakeholder expectations. Given that the dynamic environment in which your fire department operates is constantly changing, it is essential that you ensure the constant readiness of your fire department, and likewise your readiness, to address proactively the many challenges that will be faced. It is imperative that you maintain your ability to successfully manage and lead your fire department in changing and challenging times. As well as enabling you to succeed personally and professionally, the reality is that the professional development opportunities that you undertake are also instrumental in determining the success of your fire department. The value of you having a professional development plan as fire chief, and also encouraging and expecting your officers and members to have one, should be obvious.

There are many professional development opportunities available to you as a fire chief, as well as to the members of your department; however, the three major categories of professional development are training, certification, and higher education. Additional professional development opportunities that you should consider as fire chief include conferences and workshops, professional organizations and meetings, independent reading, and independent research.

Professional Development Opportunities

- Training
- Certification
- Higher Education

The *International Association of Fire Chiefs (IAFC) Officer Development Handbook* is a useful reference, as are materials available from the National Volunteer Fire Council (NVFC) and the IAFC Volunteer and Combination Officers Section (VCOS) (figure 6-1). Both of these organizations are advocates of professional development for fire chiefs and have useful professional development information available on their websites. The growing collection of applied research projects prepared by participants in the National Fire Academy's Executive Fire Officer Program and available through the National Emergency Training Center's Learning Resource Center is an extremely valuable information source that you can utilize as you seek to address challenging contemporary issues (figure 6-2).

Throughout your preparation for the position of fire chief, you participated in a number of training programs designed to prepare you in terms of the knowledge, skills, and attitudes needed to succeed in the various positions you have held within the fire department. Much of this training, particularly in the early years of your fire service career, has concentrated on technical knowledge and skills and has most likely been delivered at the local level through departmental training or local or regional fire and emergency services training academies. The methodology in much of this training has involved classroom delivery of instruction supplemented with hands-on application. Some of the courses that you likely took at the local or regional level further prepared you to be a fire officer and included such topics as strategy and tactics, incident management, instructional methodology, officership, and management.

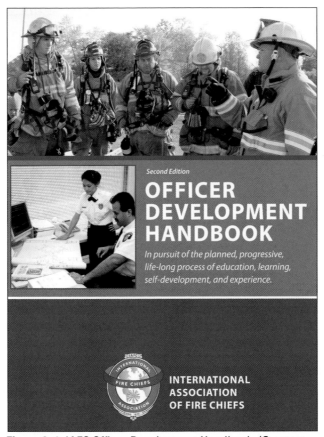

Figure 6–1. IAFC Officer Development Handbook. (Courtesy of International Association of Fire Chiefs)

Figure 6–2. NETC Learning Resource Center. (Courtesy of United States Fire Administration)

State fire training academies support and supplement the training offered at the local or regional level by offering advanced courses and training opportunities, including those specifically designed to prepare members of the fire service to advance to and through the fire officer ranks. The National Fire Academy (NFA), housed on the National Emergency Training Center (NETC) campus in Emmitsburg, Maryland, serves as the lead fire training agency in our national fire training system (figure 6–3). Most of the course offerings of the National Fire Academy are management or technical courses that complement the training courses available through local, regional, and state fire and emergency services training agencies. Resident courses are offered at the Emmitsburg campus in a number of formats, including one- and two-week courses and weekend courses. The majority of these courses have been evaluated for academic credit equivalencies by the American Council on Education (ACE). The National Fire Academy also offers regional deliveries of some courses and a growing number of online and distance delivery courses. Many of the courses offered by the Emergency Management Institute (EMI), also housed on the NETC campus, are also professional development opportunities for fire chiefs.

Figure 6–3. NETC campus. (Courtesy of United States Fire Administration)

The Executive Fire Officer Program (EFOP), the flagship program of the National Fire Academy, is a four-year program designed to prepare the next generation of fire officers. Fire chiefs from across the nation, as well as fire officers who aspire to become fire chiefs, participate in this program, which requires completion of four two-week residential courses and four associated applied research projects (figure 6-4).

Professional certification is designed to provide an independent validation of the knowledge and skills of fire and emergency services personnel. The essential role of professional certification can be further understood through clarification of a common misunderstanding regarding "training" and "certification." Certification should not be confused with the certificate that a fire service member receives after successfully completing a training course. Professional certification goes beyond training in that it utilizes written and practical skills testing based on the competencies specified in recognized professional qualifications standards, which are developed through a consensus standards-making process under the auspices of the National Fire Protection Association (NFPA) for fire service positions ranging from Firefighter through Fire Officer.

United States *Fire Administration/National Fire Academy*

Executive Fire Officer Program

Celebrating 25 Years of Excellence in Fire/Emergency
Services Executive Education: 1985–2010

 **FEMA**

Figure 6–4. NFA Executive Fire Officer Program announcement.
(Courtesy of United States Fire Administration)

The relevant standard for fire officers is the NFPA 1021—Standard for Fire Officer Professional Qualifications, which delineates required levels of knowledge and skill competencies to be certified at four levels: Fire Officer I (Supervising Fire Officer), Fire Officer II (Managing Fire Officer), Fire Officer III (Administrative Fire Officer), and Fire Officer IV (Executive Fire Officer). As a fire chief, you should desire to be certified at all of these levels. It should be recognized that professional certification, under the auspices of appropriate accrediting bodies such as the National Board on Fire Service Professional Qualifications (NBFSPQ) and the International Fire Service Accreditation Congress (IFSAC), serves as the basis for credentialing fire and emergency services personnel under a new Department of Homeland Security (DHS) initiative designed to establish a viable system for pre-qualifying emergency responders as having the appropriate knowledge and skills to perform in specified roles at major emergency incidents.

Fire Officer Certification Levels (NFPA 1021—Standard for Fire Officer Professional Qualifications)

- Fire Officer I: Supervising Fire Officer
- Fire Officer II: Managing Fire Officer
- Fire Officer III: Administrative Fire Officer
- Fire Officer IV: Executive Fire Officer

A growing number of colleges and universities offer fire science and other related academic programs that can be extremely beneficial in your professional development as a fire chief. In addition to pursuing academic degrees in the various fire and emergency services disciplines, fire chiefs and those aspiring to become fire chiefs, may also major in related disciplines, such as business management and public administration. An increasing number of these academic programs are available as blended courses, wherein some of the instruction takes place in traditional classrooms while the remainder is delivered through distance education. Many college and university courses, and in some cases entire degree programs, are now offered completely online to meet the time demands of many contemporary students, including fire service personnel.

A dilemma that many fire service personnel have faced in pursuing professional development opportunities has been an unwillingness on the part of higher education institutions to award credit for the completion of previous training or attainment of professional certifications. The Fire and Emergency Services Higher Education (FESHE) initiative facilitated by the National Fire Academy is being implemented in a growing number of states, resulting in a bridging of the gap between fire and emergency services training, certification, and higher education (figure 6-5). These initiatives are facilitating the awarding of academic credit for previous fire and emergency services training and certifications. The ACE credit recommendations for many of the NFA residential courses and the newly implemented system of awarding continuing education units for NFA courses can also enable fire service personnel to package their credentials as they pursue college degrees.

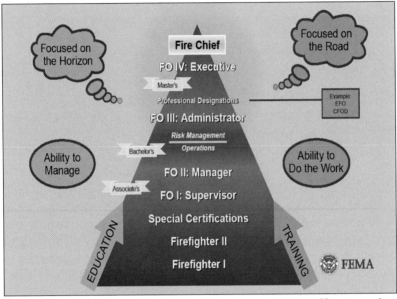

Figure 6–5. FESHE National Professional Development Model. (Courtesy of United States Fire Administration)

As a fire chief, it is incumbent upon you to be aware of all of the professional development opportunities that are available to you and to pursue selectively the opportunities that will best prepare you for continued success as a fire chief. Although you will certainly want to participate in all available relevant training courses offered at the local, regional, or state levels, avail yourself to the many courses offered through the National Fire Academy that are specifically designed, developed, and delivered to provide contemporary fire officers with the management skills, technical knowledge, and attitudes to manage and lead their fire departments successfully (figure 6-6). In addition to the outstanding classroom instruction that you will receive, you will likewise benefit from the networking opportunities of attending these courses. Pursue relevant professional certifications, including Fire Officer certification.

Figure 6–6. National Fire Academy course catalog. (Courtesy of United States Fire Administration)

Last, but certainly not least, avail yourself of the opportunities available through colleges and universities to enhance your knowledge and skills through academic preparation. The growing ability to leverage your previous training and professional certifications as you pursue an academic degree, as well as the growing number of course delivery alternatives, will enable you to do so effectively and efficiently while you continue to attend to your daily responsibilities as fire chief.

Job Aid 6 that appears at the end of this lesson will assist you in developing a realistic professional development plan designed to contribute to your professional and personal success, as well as the success of the fire departments that you will manage and lead throughout your career.

Things to DO:

- Recognize that your success as a fire chief and that of your fire department will be influenced by your ability to maintain and enhance your knowledge and skills.

- Recognize the importance of continuous professional development in maintaining and enhancing your ability to successfully manage and lead your fire department.

- Develop a professional development plan for yourself, and encourage others in your fire department to do likewise.

- Recognize that, in addition to the personal and professional benefits that you will realize through participating in professional development opportunities, your fire department will also benefit from your engaging in these activities.

- Become familiar with and take advantage of professional development opportunities and resources available through national fire service organizations, including the *International Association of Fire Chiefs (IAFC) Officer Development Handbook.*

- Participate in training offered within your fire department, as well as through local, regional, and state fire and emergency services training academies.

- Become familiar with and participate in relevant residential, regional delivery, and online courses offered by the National Fire Academy and the Emergency Management Institute.

- Consider participating in the National Fire Academy's Executive Fire Officer Program.

- Pursue networking opportunities through participation in training courses, conferences, and professional organizations.

- Pursue appropriate professional certifications, including those as a fire officer and in other appropriate specialties based on the scope of operations of your fire department.

- Pursue appropriate coursework and academic programs offered by colleges and universities.

Things NOT to Do:

- Rationalize that as fire chief you do not have the time or need to pursue professional development opportunities.
- Fail to recognize that the dynamic environment in which your fire department operates will require that you maintain and enhance your knowledge and skills.
- Fail to utilize information resources such as the *Executive Fire Officer Program* Applied Research Projects.
- Fail to encourage other members of your fire department, particularly those who currently serve as fire officers or aspire to do so, to engage in available professional development opportunities.

Developing a Professional Development Plan (Job Aid 6)

1. Identify your professional goals for the next five years with respect to training.

2. Identify your professional goals for the next five years with respect to professional certifications.

3. Identify your professional goals for the next five years with respect to higher education.

4. Identify your professional aspirations for the next five years.

Lesson 7: Management Skills You Will Need

Role in Survival and Success

As you advance through the levels of management within an organization, the management skills necessary to successfully enact your roles and responsibilities change in accordance with the management level of each respective position. As you make the transition from line officer to chief officer and later to the position of fire chief, you are advancing in terms of both rank and management level within the fire department. Accompanying your advancement within the fire department is an increase in the scope of your responsibilities, the time horizon of your decision making, and the primary management skills that you will utilize.

These management skills can be categorized into three sets of skills: technical skills, human skills, and conceptual skills. As you assume the position of fire chief, your responsibilities will require that you less frequently utilize your technical skills, while more routinely engaging in the use of conceptual skills. Human skills will also be crucial to your success and survival as a fire chief responsible for the successful management and leadership of your fire department.

What You Need to Know

Managers employ three sets of management skills as they enact their roles and responsibilities within an organization: technical skills, human skills, and conceptual skills.

Management Skills

- Technical skills
- Human skills
- Conceptual skills

Technical skills include knowledge and skills in a particular discipline or field, such as firefighting or vehicle rescue. Human skills involve the ability to work or interact with others. Conceptual skills involve the ability to conceptualize abstract and complex situations, as well as to see the "big picture."

Technical Skills

Knowledge and skills in a particular field or discipline

Human Skills

Skills that relate to working with people

Conceptual Skills

Skills that involve being able to see the big picture

Three levels of management typically exist within an organization. In ascending order, these three management levels are: the operational level, the tactical level, and the strategic level. These management levels can be differentiated based on the managers involved, their scope of decision making, the time horizon of decision making, and the primary management skills used.

Levels of Management

- Operational (functional)
- Tactical (business)
- Strategic (corporate)

The operational level is also known as the functional level of an organization. It is at this organizational level that first-level managers, such as a lieutenant or captain, make short-term decisions that affect only their unit. The primary management skills used at this management level are technical skills. Human skills are also important at this and the other levels of management, whereas the use of conceptual skills is infrequent at this management level. A lieutenant or captain preparing a training schedule for his or her station for the next three months would be an example of this level of management.

The tactical level, also called the business level, comprises the middle level of management within an organization. Middle managers make intermediate-term decisions that impact their divisions, such as when a battalion chief coordinates the annual fire prevention and public education programs for his or her assigned fire stations. The primary management skills used at this level are human skills.

The senior managers of an organization function at the strategic level, which is also referred to as the corporate level. The primary skills used by this level of managers are conceptual skills. Human skills are also important in enacting the roles and responsibilities of this management level. Managers at this level, including the fire chief and his or her senior staff, engage in activities designed to prepare and position the fire department for present and future success.

As you progressed through the management ranks to the position of fire chief, you most likely realized the importance of human skills to your success in each position and at each management level. The importance of human skills holds true as you seek to enact your management and leadership responsibilities as fire chief. Although it is still important for you to have a working knowledge of the technical operations of your fire department, you will utilize your technical skills much less frequently and your conceptual skills more routinely as you serve as fire chief. This is due to the fact that your roles and responsibilities as fire chief typically involve a decrease in technical activities and involvement, with a corresponding increase in management and leadership activities, such as facilitating organizational strategic planning processes, developing and managing budgets, and initiating and managing needed change within your fire department. As the fire chief, you must resist the temptation to continue to perform technical tasks, rather than delegating these tasks to the appropriate subordinates. Not doing so can significantly compromise your success as fire chief as well as that of your fire department.

Job Aid 7 that appears at the end of this lesson will assist you in identifying and considering the management skills that will be required to enable you to succeed as a fire chief and to prepare your fire department for present and future success.

Things to DO:

- Understand the three types of management skills and their applicability to the different management levels within your fire department.

- Recognize that as fire chief your responsibilities have expanded in scope, as has the time horizon of your decision making.

- Recognize the importance of human skills to the successful enactment of your management and leadership responsibilities and corresponding managerial roles as fire chief.

- Recognize that the primary management skills you will utilize as fire chief will be conceptual skills.

- Delegate responsibilities and tasks to subordinates when appropriate.

Things NOT to Do:

- Assume that the management skills required to succeed in your previous positions will enable you to succeed as fire chief.

- Engage in technical operations and activities that should be the responsibilities of your subordinates.

- Fail to prepare and position your fire department to effectively, efficiently, and safely achieve its mission and meet and, where possible, exceed the realistic expectations of its stakeholders.

Necessary Management Skills
for the Fire Chief
(Job Aid 7)

1. Describe how technical skills will contribute to your success as fire chief.

2. Describe how human skills will contribute to your success as fire chief.

3. Describe how conceptual skills will contribute to your success as fire chief.

4. Identify the challenges that you will face in enacting technical, human, and/or conceptual skills as fire chief.

Lesson 8: Delegation is Essential

Role in Survival and Success

One of the greatest challenges that you will face as a fire chief is finding the time to enact all of your roles and responsibilities, both inside and outside your fire department. As a new fire chief, it is easy to convince yourself that you have to do everything yourself. It is imperative that you avoid adopting such a mindset in that it will undermine your success as fire chief and that of your fire department, as well as limit the opportunities for other members of your fire department to fully utilize their talents and further enhance their own knowledge and skills.

Time is a limited resource that you, as fire chief, must learn how to manage properly. Time management is a crucial management survival strategy at which you need to become proficient. An essential aspect of effective time management is the use of delegation within your fire department. Through delegation, you will be able to assign certain responsibilities and tasks to your fire officers and other members of your fire department in the interest of enhancing the overall effectiveness of your fire department and the efficiency of its use of resources, including personnel. The effective use of delegation will additionally free up your time to focus on key roles and responsibilities related to the successful management and leadership of your fire department, and will prepare and position your department for present and future success. Through time you will come to appreciate that the use of effective delegation is an essential skill of the contemporary fire chief and that it plays a significant role in professional, as well as organizational, success.

What You Need to Know

There are four basic ways to accomplish work within your fire department: retain the work yourself, delegate responsibility for the work, delegate decision-making authority, or refer the task. All too often fire chiefs, particularly those new to the position, assume that they must do everything themselves and thus retain many responsibilities and tasks that should rightfully be delegated to others within their fire departments. The use of delegation is essential in that it enhances the effectiveness and efficiency of accomplishing the work of the fire department, as well as contributing to the development of subordinates and preparing them for future responsibilities within the department. Effective delegation can produce beneficial outcomes for you as fire chief, for the subordinates to whom you delegate, and for your fire department.

Basic Ways to Accomplish Work

- Retain work
- Delegate responsibility for work
- Delegate decision-making authority
- Refer task

Beneficiaries of Effective Delegation

- The manager
- The subordinate
- The organization

Delegation is a process through which you can increase your personal effectiveness, as well as that of your fire department (figure 8–1). The delegation process involves passing responsibility for tasks, as well as the accompanying authority to accomplish these tasks, to subordinates within your fire department. It is important that the appropriate and necessary authority to successfully enact the delegated responsibility be provided to the subordinate accepting the delegation. An essential third component of effective delegation is the establishment of accountability with respect to performing the assigned responsibility through the use of the delegated authority.

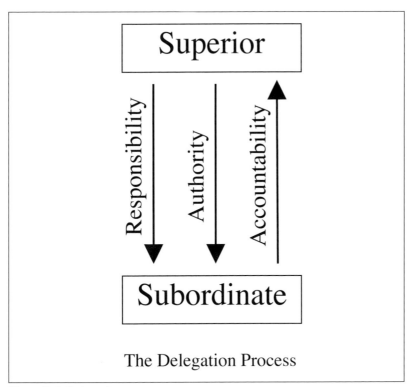

The Delegation Process

Figure 8–1. Delegation process

Delegation can be accomplished through orders and directives, as well as through inclusion of specific responsibilities in one's job description. Examples of these types of delegation would be the assignment of a fire officer to handle operations at an emergency incident or the appointment of a departmental training officer. (See figures 8-2 and 8-3.)

Figure 8–2. Fire chief making fireground assignment. (Courtesy of Broomall Fire Company)

Figure 8–3. Fire chief discussing assignment with subordinates

An essential aspect of successful delegation is the selection of appropriate individuals who are qualified or qualifiable to accept the task or responsibility and likewise have the interest and willingness to do so. Delegation is appropriate in situations where subordinates have the necessary training or expertise and there is sufficient time available to delegate. In addition to being a crucial management survival skill for you as fire chief, the use of effective delegation provides professional development opportunities for the members of your fire department.

Selection of Delegation Recipients

- Individuals who are qualified or qualifiable to complete the assignment
- Individuals who are interested and willing to accept the assignment

Whereas at times some fire chiefs are reluctant to delegate for a variety of reasons, which results in compromising their success and that of their fire department, it is imperative to understand that there are those roles and responsibilities that are inherent to the position of fire chief and should, therefore, not be delegated to others. Possible obstacles to effective delegation include: the manager's reluctance to delegate, the subordinate's reluctance to accept the delegated task or responsibility, and organizational factors that frustrate or discourage the use of delegation or its effectiveness.

Potential Obstacles to Effective Delegation

- Manager's reluctance to delegate
- Subordinate's reluctance to accept delegation
- Organizational factors that discourage delegation

Your professional success as a fire chief and that of your fire department will in large part be determined by your willingness and skill in the effective use of delegation. Should you fail to delegate to others when appropriate, you will be compromising your success and needlessly complicating your job by allocating your limited time to matters and activities that do not require your direct attention and involvement. Don't make the mistake of assuming that as fire chief you must know all and be capable of doing all. Your success as fire chief, as well as that of your fire department, will be greatly enhanced when you surround yourself with talented and committed people and develop the confidence to trust, motivate, and empower them through the effective use of delegation.

Job Aid 8 that appears at the end of this lesson will assist you in evaluating and improving your use of delegation, as well as the use of delegation within your fire department.

Things to DO:

- Resist the tendency to feel that you have to do everything yourself, and be directly involved in all the aspects and activities of your fire department.

- Recognize the value of the use of effective delegation to you as fire chief, to your subordinates, and to your fire department.

- Recognize that through effective delegation you can enhance your effectiveness and success, as well as that of your fire department.

- Use effective delegation as a professional development strategy for members of your fire department.

- Delegate to individuals who are qualified or qualifiable, as well as interested and willing to accept the delegation.

- Make delegation assignments clear in terms of responsibility, authority, and accountability.

- Inform all involved personnel of delegation assignments.

- Hold those receiving delegated assignments accountable for performance.

- Surround yourself with competent people and demonstrate your confidence and trust in them through making delegation assignments that contribute to their motivation, empowerment, and professional development.

Things NOT to Do:

- Fail to recognize the significant challenges that you will face as fire chief in allocating your limited time to the many conflicting time demands of the position of fire chief.

- Fail to provide the necessary and appropriate authority to accomplish delegated responsibilities.

- Delegate responsibilities or tasks that should be retained by you as fire chief or that are unlikely to be successfully performed by others.

Enhancing Delegation within Your Fire Department (Job Aid 8)

1. Describe the present use and effectiveness of delegation within your fire department.

2. Identify the factors that currently contribute to the effective or ineffective use of delegation within your fire department.

3. Discuss your ability and comfort level with respect to delegating assignments and responsibilities to others within your fire department.

4. Identify strategies you could use to enhance your personal effectiveness in delegation, as well as that of your fire department.

Lesson 9: Positioning Your Department for Success

Role in Survival and Success

As fire chief, you play a crucial role in preparing and positioning your fire department for both present and future success. This role involves charting the course for your organization and gaining the support of fire department members and other stakeholders in support of that desired destination and route of travel. Throughout your service as fire chief, you will have the opportunity and responsibility to articulate a vision for your fire department and to prepare and position your department to attain that desired future state. Your conceptual skills and visionary leadership will be instrumental in preparing and positioning your fire department to succeed in the present and the future.

You will quickly discover, if you have not done so already, that the success of an organization does not "just happen"—you need to plan for and work to achieve success. Your visionary leadership and astute management skills will be needed to prepare and position your fire department to respond to current issues and challenges, as well as to anticipate and prepare to address future events and challenges. As the fire chief, you have a responsibility to incorporate stewardship as you manage and lead your fire department in a manner that prepares it to survive and succeed in meeting and, where possible, exceeding realistic stakeholder expectations in an effective, efficient, and safe manner. Through the use of environmental scanning and strategic planning, you will have the opportunity to create a successful future for your fire department and the stakeholders who depend upon it to deliver essential emergency services.

What You Need to Know

There are two distinct approaches available to you and your fire department with respect to preparing for the future. You can adopt a reactive approach wherein you do not plan for or concern yourself with the future and simply accept and deal with future challenges and developments as they present themselves. A more realistic approach for you and your fire department to adopt is to be proactive in terms of developing a thorough understanding of your fire department's strengths and weaknesses, in light of the environmental opportunities and threats that it faces, and use this understanding to provide the necessary insights to develop and implement a realistic strategic plan for the future, designed to prepare and position your department for success and survival. The merit of a proactive approach should be obvious in terms of not leaving things in the hands of fate and enabling your fire department to pursue appropriate strategic opportunities that enhance its ability to more fully meet and, where possible, exceed the reasonable expectations of its stakeholders.

Organizational Planning Approaches

- Proactive
- Reactive

The logical starting point in preparing and positioning your fire department for the future is to develop and gain the support of internal and external stakeholders for a mission statement that delineates why the organization exists and provides invaluable and essential strategic direction for those with responsibilities for managing and leading the department. This mission statement should incorporate stakeholder input and expectations and serve as the basis for operating as a mission-driven organization, meaning that the mission of the organization drives everything that your fire department does. An example of this might be an expansion of public education programs in response to inclusion in the fire department's mission statement of a commitment to community risk reduction.

Mission

A statement of an organization's purpose or reason for being

As you plan for the future in the interest of preparing and positioning your fire department, it is imperative that you fully understand and consider the various dimensions of your department's environment that have a bearing on its future in terms of presenting challenges and issues which you will need to address in a proactive manner. The external environment of your fire department is composed of economic, political, legal, social, cultural, demographic, and technological dimensions. Although it would be an impossible task for you, as fire chief, to attempt to monitor and fully understand all aspects of this external environment, it is in your best interest to understand general environmental trends and elements of the task environment that are specifically relevant to and have the potential to impact your fire department. In maintaining a current awareness of the environment, you will find its dynamic nature challenging; it is constantly changing and continually presenting new challenges for you and your fire department to address.

External Environment

- General environment
- Task environment

You will discover that the economic environment will present many challenges throughout your career as fire chief, including resource and funding uncertainties, budget unpredictability, increased capital and operating costs, reduced community support, downsizing, and pressure to consider consolidation, redistricting, or regionalization. The political dimension will become an ever-present reality for your fire department and for you as fire chief. You will come to appreciate the significant impact that existing and proposed laws, regulations, and standards can have on your department. Although it is sometimes easy to understate their importance, sociocultural and demographic changes have had pronounced impacts in many areas, including the demand for fire and emergency services, stakeholder expectations, and community and departmental diversity. A prime example of the impact of demographic changes is the increased demand for emergency medical and other fire department services resulting from the aging nature of the population. Last, but certainly not least, is the technological environment, which impacts many aspects of the operations of the contemporary fire department both on and off the incident scene.

Environmental Dimensions

- Cultural
- Demographic
- Economic
- Legal
- Political
- Social
- Technological

As the fire chief, proactively manage and lead your fire department through the use of a strategic planning process designed to ascertain your department's present and likely future situation and to develop and implement strategies designed to enable the department to accomplish goals consistent with its mission. Facilitate a planning process that affords organizational stakeholders, including fire department members, the opportunity to be involved and provide input (figure 9–1).

Figure 9–1. Fire department strategic planning session

It is also important that you use an environmental scanning approach that seeks to identify the organizational strengths and weaknesses of your fire department, keeping in mind the environmental opportunities and threats that it faces in the present and will likely face in the future. The understanding and insights gained through environmental scanning enable you to make informed decisions regarding an appropriate strategic direction to pursue, along with the realistic goals and strategies to achieve that desired future state. Planning in this manner is proactive, whereas planning without the guidance derived from conducting an environmental analysis would be reactive.

SWOT Analysis

- Organizational strengths
- Organizational weaknesses
- Environmental opportunities
- Environmental threats

Organizational strengths include attributes or distinctions of the organization that can potentially contribute to its success, whereas organizational weaknesses consist of the limitations of the organization, which can contribute to potential harm and compromise success. In considering the potential strengths and weaknesses of your fire department, take into account apparatus, equipment, facilities, financial situation, human resources, internal climate and morale, leadership, and response capacity and readiness, as well as other relevant organizational factors.

Potential Organizational Strengths and Weaknesses

- Apparatus and equipment
- Facilities
- Financial situation
- Human resources
- Internal climate and morale
- Leadership
- Response capacity
- Response preparedness (readiness)
- Training and knowledge
- Other organizational factors

Environmental opportunities are positive situations in the environment from which the organization can potentially benefit, whereas environmental threats are negative situations in the environment that can compromise organizational success and potentially harm the organization. Among the factors that you will want to consider in evaluating environmental opportunities and threats are: changing laws, regulations, and standards; changing social and cultural norms; economic developments; fundraising potential; municipal support; new initiatives; new technologies; political developments; consolidation, regionalization, and redistricting; and other relevant environmental elements.

Potential Environmental Opportunities and Threats

- Changing laws, regulations, and standards
- Changing social and cultural norms
- Economic developments
- Fundraising potential
- Municipal support
- New initiatives
- Political developments
- Redistricting/regionalization/consolidation
- Other environmental factors

As you facilitate the planning process within your fire department, ensure that informed decisions are made based on the understanding and insights developed through environmental analysis, and that the resulting strategic plan is developed so as to build on organizational strengths and minimize organizational weaknesses, while also taking advantage of environmental opportunities and avoiding environmental threats. Through strategic planning, you can leverage this knowledge to formulate and successfully implement the necessary strategies to achieve a realistic set of organizational goals that are consistent with your fire department's mission and the expectations of its stakeholders.

Your fire department will benefit from the utilization of a comprehensive strategic planning approach or process that incorporates three sequential phases: a preparation phase, a planning phase, and an implementation phase. During the preparation phase, your fire department should make a commitment to the strategic planning process, organize its planning resources, form a strategic planning committee or team, and establish advisory groups to assist in the planning process.

Strategic Planning Process

1. Preparation phase
2. Planning phase
3. Implementation phase

Preparation Phase

- Make planning commitment
- Organize planning resources
- Form planning team
- Establish advisory groups

The second phase in the strategic planning process, the planning phase, begins with determining information needs and the collection and analysis of necessary data. It is during this phase within the strategic planning process that a mission is articulated, which provides the necessary strategic direction for the formulation of goals and objectives. The last step in this strategic planning phase is to identify appropriate strategies to support these goals and objectives.

Planning Phase

- Determine information needs
- Collect and analyze data
- Articulate a mission
- Formulate goals and objectives
- Identify appropriate strategies

Goals serve to identify a precise and measurable desired future state that the organization attempts to realize. Well-written goals provide essential strategic direction in the development and implementation of a strategic plan. Goals should be written so as to be measurable and time-specific. They should be both challenging and realistic. An example of a fire department goal statement is:

Increase average personnel response for emergency calls from 12 members to 15 members within the next three years.

Goal

An intended outcome that an organization seeks to achieve

Characteristics of Goals

- Measurable
- Time-specific
- Challenging
- Realistic

Objectives measure progress towards goal attainment in two ways. Program objectives measure progress in implementing strategies, whereas impact objectives measure progress towards actual goal attainment. Representative program objectives associated with the previous goal include:

Year 1:
Develop and implement a new recruitment program.

Provide enhanced training opportunities for new and current members.

Year 2:
Develop and implement a new rewards and incentives program.

Provide enhanced training opportunities for new and current members.

Year 3:
Evaluate recruitment and retention initiatives.

Revise recruitment and retention program and initiatives as necessary.

The corresponding impact objectives for the above program objectives include:

Year 1:
Increase average response for emergency calls to 13 members.

Year 2:
Increase average response for emergency calls to 14 members.

Year 3:
Increase average response for emergency calls to 15 members.

Types of Objectives

- Program objectives
- Impact objectives

Program Objective

A measure of progress in the implementation of strategies

Impact Objective

A measure of progress towards goal attainment

Strategies are the means or actions through which an organization attempts to achieve a stated goal. Representative strategies associated with the previous example are:

Develop and implement a new recruitment program.

Provide enhanced training opportunities for new and current members.

Develop and implement a new rewards and incentives program.

Evaluate recruitment and retention initiatives.

Revise recruitment and retention program and initiatives as necessary.

Strategy

An action or means taken to achieve an end or goal

The third and final phase of strategic planning is the implementation phase. During this phase, it is important to gain support for the strategic plan that has been developed and properly implement the plan. The periodic evaluation and, as necessary, revision of the plan represents the final activity in the strategic planning process.

Implementation Phase

- Gain support for plan
- Implement plan
- Evaluate and, as necessary, revise plan

It is crucial that you realize the importance of facilitating your department's strategic planning process in a manner that leads to those who will be required to implement the resulting plan having ownership in it and thus commitment to work towards its successful implementation. In addition to your visionary leadership and facilitation skills, the successful development and implementation of your fire department's strategic plan will often require you to utilize your skills as a change agent in assisting fire department members and other organizational stakeholders in overcoming their resistance to change.

Job Aids 9A and 9B that appear at the end of this lesson are designed to enable you to conduct an organizational and environmental analysis of your fire department and to facilitate the development of realistic goals, objectives, and strategies for your fire department.

Things to DO:

- Recognize the instrumental role that you, as fire chief, play in preparing and positioning your fire department for present and future success.

- Recognize your responsibility to demonstrate stewardship as you manage and lead your fire department.

- Adopt and institute a proactive approach to planning within your fire department.

- Monitor relevant aspects of the general and task environments of your fire department and use this information to inform your decision making and planning activities.

- Facilitate the development of a realistic mission statement that departmental stakeholders are prepared to endorse and support.

- Utilize environmental scanning to identify the organizational strengths and weaknesses of your fire department and the opportunities and threats present in its environment.

- Incorporate the understanding and insights gained through environmental scanning in the development of your strategic plan.

- Develop a strategic plan that builds on organizational strengths, minimizes organizational weaknesses, pursues environmental opportunities, and avoids environmental threats.

- Ensure that the goals and objectives articulated in your strategic plan are consistent with your fire department's mission.

- Incorporate relevant stakeholder expectations and input into your strategic plan.

- As necessary, serve as a change agent and assist organizational stakeholders in overcoming resistance to change related to the planned initiatives articulated in your department's strategic plan.

Things NOT to Do:

- Fail to articulate a realistic vision for your fire department that provides essential strategic direction.

- Engage in reactive, rather than proactive, planning.

- Utilize a strategic planning process that does not provide sufficient opportunities for stakeholder involvement and input.
- Fail to ensure that those who will be responsible for implementing the strategic plan are involved in the planning process to the extent that they support the plan and are committed to work towards its successful implementation.

Conducting a SWOT Analysis
(Job Aid 9A)

1. Identify the present organizational strengths of your fire department.

2. Identify the present organizational weaknesses of your fire department.

3. Identify the present and likely future environmental opportunities that your fire department faces.

4. Identify the present and likely future environmental threats that your fire department faces.

Formulating Goals, Objectives, and Strategies (Job Aid 9B)

1. Identify a goal that you would like to pursue for your fire department over the next few years.

2. Identify the program objectives and impact objectives associated with the previous goal.

3. Identify the strategies necessary to achieve the previous goal.

4. Discuss the potential implementation issues in using the previous strategies to achieve the stated goal.

Lesson 10: Understanding and Meeting Stakeholder Expectations

Role in Survival and Success

Your success as fire chief, as well as the success of your fire department, will in large part be determined by your understanding of the stakeholders of your fire department and their expectations. Organizational stakeholders are individuals, groups, and organizations that are directly or indirectly influenced by the goals that an organization pursues and its success in achieving them. Stakeholders thus have a vested interest in what your fire department attempts to accomplish and its success towards that end.

> **Stakeholders**
>
> Individuals, groups, or organizations with an interest in and expectations for an organization

As the fire chief, you have a responsibility to understand the expectations of your fire department's stakeholders and prepare and position the department to meet and, where possible, exceed their reasonable expectations. An integral aspect of your role with respect to stakeholders may be assisting them in developing reasonable expectations, as in the case of realistic response times for an unmanned volunteer fire station. Your professional success and that of your fire department will require that you fully understand and respond to the realistic expectations of the stakeholders of your fire department.

What You Need to Know

The stakeholders of your fire department fall into two categories: internal stakeholders and external stakeholders. Internal stakeholders include fire department members, elected and appointed officials, and other fire and emergency services organizations and agencies. Individuals who live in, work in, or travel to or through your jurisdiction, as well as organizations located or operating within your response territory, are considered external stakeholders. Although it is unlikely that you, as a fire chief, would fail to consider external stakeholders of your fire department and their expectations, a common oversight on the part of fire chiefs is to overlook the department's internal stakeholders, particularly its members.

Organizational Stakeholders

- Internal stakeholders
- External stakeholders

Your fire department will face a number of challenges in service delivery. These challenges include the fact that it is extremely difficult to predict with any certainty when requests for service will be received, resulting in unscheduled service delivery and varying call volumes. Many of the incidents to which your fire department will respond require time-critical services with immediate service consumption. These services are typically labor-intensive in terms of human resource requirements.

It is imperative that you, as fire chief, and the department that you manage and lead understand the expectations of organizational stakeholders in order to prepare and position your department to meet and, where possible, exceed their reasonable expectations. Stakeholder analysis, wherein you identify the various relevant stakeholders along with the expectations of each group, will equip you with the understanding necessary to prepare and position your department to successfully respond to these expectations. Although most stakeholder expectations will prove to be reasonable and realistic, some will not and will require you, through the use of communication and education, to assist these stakeholders in adjusting their expectations so as to be reasonable and within the reach of your fire department and its resources and capabilities.

Stakeholder Analysis

Process of identifying organizational stakeholder groups and developing an understanding of the expectations of each group

The typical expectations that stakeholders have with respect to their fire department and the services that it provides fall into the following categories: accessibility, completeness, consistency, convenience, courtesy, effectiveness, efficiency, image, professionalism, responsiveness, safety, and timeliness. Stakeholders expect that when they request a fire department response to an emergency, the fire department will be available to respond. They expect that the fire department will deliver the comprehensive services necessary to fully resolve an emergency situation and that, regardless of the time or day of the response, consistent services will be delivered. Stakeholders also expect that the process for requesting fire department services will be easy and understandable and also that fire department personnel will be courteous as they enact their responsibilities.

Fire department stakeholders expect that the fire department will operate effectively and efficiently and that is will exhibit stewardship in its use of resources. They expect that the fire department will portray an appropriate image, with fire department personnel conducting themselves as highly trained professionals at all times. Stakeholders expect that the fire department and its personnel will be responsive to their emergency situation and its needs and will utilize the necessary and appropriate strategies and tactics to ensure the life safety of both response personnel and the public. Last, but certainly not least, stakeholders expect that the fire department will, upon being dispatched to an emergency call, respond, arrive, and resolve the emergency situation in a timely manner.

Stakeholder Expectations for Fire and Emergency Services

- Accessibility
- Completeness
- Consistency
- Convenience
- Courtesy
- Effectiveness
- Efficiency
- Image
- Professionalism
- Responsiveness
- Safety
- Timeliness

As an astute fire chief, you will quickly develop an appreciation of the role of perception versus reality in the time of an emergency. To the resident who calls the fire department to respond to the "one room and contents" kitchen fire, it may seem to take a rather long time for the fire department to arrive on the scene, when in reality the fire department made the response and arrived in a reasonable time.

Your role as fire chief with respect to stakeholder expectations is to ensure that all relevant stakeholder groups are identified, that their expectations are understood and realistic, and that your fire department is prepared to meet and, where possible, exceed reasonable stakeholder expectations (figure 10–1). The use of strategic planning will enable you to properly prepare and position your department, as well as yourself as fire chief, to successfully address the expectations of your department's internal and external stakeholders.

Figure 10–1. Fire chief meeting with stakeholder group. (Courtesy of Alison Miller)

Job Aid 10 that appears at the end of this lesson is designed to assist you in identifying the various internal and external stakeholders of your fire department, as well as the expectations that each group has for your department.

Things to DO:

- Recognize the importance of understanding stakeholders and their expectations for your fire department.
- Recognize and enact your role as fire chief with respect to preparing your fire department to understand and also meet and, where possible, exceed reasonable stakeholder expectations.
- Utilize stakeholder analysis to develop a full understanding of the internal and external stakeholders of your fire department and the expectations of each stakeholder group.
- Prepare and position your fire department to meet and, where possible, exceed reasonable stakeholder expectations.
- Understand the unique challenges involved in delivering fire department services.

Things **NOT** to Do:

- Fail to understand the typical expectations that stakeholders have for their fire department.
- Fail to assist organizational stakeholders in developing realistic and reasonable expectations for your fire department.
- Fail to incorporate stakeholder expectations in strategic planning activities.

Identifying Stakeholders and Their Expectations (Job Aid 10)

1. Identify the internal stakeholder groups with respect to your fire department.

2. Identify the external stakeholder groups with respect to your fire department.

3. Identify the expectations of each internal and external stakeholder group.

4. Describe the challenges of meeting the expectations of each stakeholder group.

Lesson 11: Recruiting and Retaining Personnel

Role in Survival and Success

Recruitment and retention of fire department personnel will be one of your most important and challenging responsibilities as fire chief. As a fire chief, you will discover that human resources are the most important and essential resource of your fire department and a major determinant of your department's success in achieving its mission and delivering the essential emergency services that its stakeholders expect and deserve. Your success as fire chief, and that of your fire department, requires the recruitment and retention of the necessary cadre of personnel who possess the requisite knowledge, skills, and attitudes to effectively, efficiently, and safely enact their assigned roles and responsibilities both on and off the incident scene.

As the fire chief, you are responsible for ensuring your department's readiness to respond to and address the fire and life safety needs of the community that your department serves. Your success in the mission-critical areas of personnel recruitment and retention requires you to understand the challenges that your fire department faces with respect to recruiting and retaining personnel and to formulate and implement appropriate recruitment and retention strategies, as well as related strategies designed to motivate and empower the members of your department. Your success in this essential area will have a marked impact on your professional success as fire chief and the success of your fire department.

What You Need to Know

To operate your fire department in an effective, efficient, and safe manner that is responsive to its mission and the expectations of its stakeholders, you will need a sufficient number of highly qualified and motivated personnel who possess the necessary knowledge, skills, and attitudes to perform their specific roles and responsibilities and thus accomplish the work of the fire department. Three primary staffing approaches exist within contemporary fire departments and departments are thus classified as volunteer, paid, or combination. In a volunteer department, all personnel serve their community without being paid for their service. Personnel serving in a paid department are compensated for their service though wages, salaries, and/or benefits. In a combination department, some personnel volunteer their time and others are compensated. Recruiting qualified and motivated personnel has become a serious problem in a growing number of fire departments, particularly in those that rely on volunteers for staffing, given that attracting volunteers has become increasingly difficult in many communities.

Fire Department Staffing Models

- Volunteer department
- Combination department
- Paid department

Personnel Recruitment

Attracting interested individuals and selecting new personnel to serve in the fire department

The successful recruitment of fire department personnel begins with defining the responsibilities of the various positions within the fire department and the necessary qualifications, in terms of knowledge, skills, and attitudes, to succeed in each position. This information should be documented in a job description and job specification, respectively.

Job Description

A written document that lists the duties, responsibilities, and working relationships of a job

Job Specification

A written document that delineates the necessary qualifications to successfully perform a job

It is also essential that you develop an understanding of why members desire to join your fire department, whether as volunteers or in a paid capacity. You will find that there are numerous general reasons why individuals decide to join a fire department and, in some cases, specific reasons associated with that department. A frequent reason why individuals join a fire department as volunteers is to serve their community and help others in their time of need. One of the obvious reasons that individuals accept paid positions within a fire department is to earn a living and support themselves and their families. It is interesting to note that there are some reasons that result in the successful recruitment of volunteers and paid personnel alike.

The success of your department's recruitment activities will be determined by your understanding of the reasons why members join your department, the resulting expectations of these members with respect to your fire department, and what your fire department needs to do to meet and, where possible, exceed the reasonable expectations of fire department members. Develop and implement a personnel recruitment program tailored to the specific needs and challenges of your fire department that ensures that potential applicants are afforded a realistic preview of your department and the roles and responsibilities of its members through this program.

Recruitment Program

An organization's approach to recruiting internal and/or external job candidates

Realistic Job Preview

Providing job candidates with an accurate understanding of a job and its responsibilities and expectations

Analyze the traditional sources of applicants who apply to join your fire department along with the yield of each of the recruitment sources that your department has traditionally employed. Where appropriate, expand the scope of your recruitment activities in terms of your recruitment target audience and the strategies and means you are using to reach potential candidates. Although the typical approaches used in recruiting fire department personnel include recruitment announcements, brochures, posters, signs, newspapers, radio, and television, you will find that an attractive, informative, and easy-to-navigate website is a valuable tool in attracting new members (figure 11–1).

Although it is commonplace within the fire service to refer to "personnel recruitment," there are actually two distinct human resource management activities involved in bringing new members into your fire department, or "on the job," as this is commonly called. These activities are recruitment and selection. Whereas recruitment involves attracting interested applicants or candidates to join your fire department, selection involves determining which interested individuals should be offered the opportunity to affiliate with your department.

In conducting these recruitment and selection activities, it is imperative that they be done in compliance with all relevant laws and regulations, whether personnel are joining the department as volunteers or in paid positions. State mandates, consent decree mandates, and court decisions may have standing in certain situations. As the fire chief, you must become knowledgeable in these requirements and act in conformance with them in your department's recruitment and selection activities. Failure to do so can result in serious consequences for both you as fire chief and for your fire department. The merit of seeking counsel and discernment from human resource management and legal professionals should be recognized.

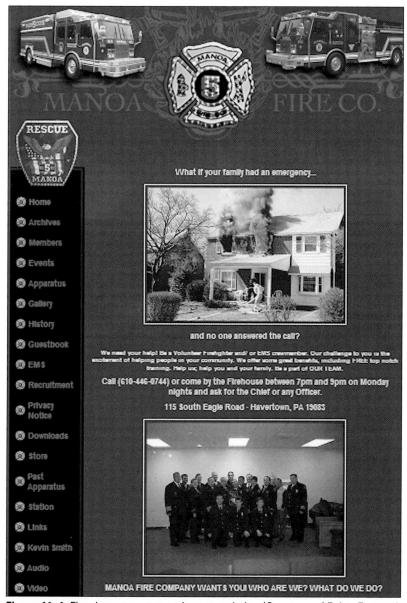

Figure 11–1. Fire department recruitment website. (Courtesy of Brian Feeney/ Manoa Fire Company)

A second, and equally challenging, human resource issue that you and your fire department may face is the retention of fire department personnel. Many contemporary fire departments continually struggle with the issues of recruitment and retention. Some, however, experience significant success in the recruitment of personnel only to have difficulty later in retaining these same individuals.

> **Personnel Retention**
>
> The desire to have members continue their affiliation with the fire department

The factors that contribute to personnel retention often correspond with the reasons individuals join the fire department in the first place and the expectations they have in so doing. As fire chief, it is crucial that you seek to fully understand these reasons and to create a work environment that addresses them appropriately. Unless you understand and define your department's retention problem and its contributing factors, you will not be likely to succeed in addressing it and improving personnel retention. Providing a realistic preview of the fire department and what it expects of its members during the recruitment process is essential from the standpoint of enhancing the likelihood of subsequent member retention.

You will find that personnel recruitment and retention are companion activities and that the issues and challenges associated with each are interrelated. The success of your fire department in these critical areas will derive from the adoption of a proactive approach. The issues of recruitment and retention are too important to allow them to be addressed in a complacent or reactive manner. It will be through your proactive and visionary leadership and management approach, supported by the officers and members of your department, that your department will be able to succeed in the recruitment and retention of the cadre of highly qualified personnel necessary to effectively, efficiently, and safely achieve its mission and goals, while meeting and, where possible, exceeding the reasonable expectations of its stakeholders.

Job Aid 11 at the end of this lesson is designed to assist you in developing an understanding of the personnel recruitment and retention challenges within your fire department and in identifying strategies that could be used to enhance the success of your department's personnel recruitment and retention.

Things to DO:

- Recognize the importance of effective recruitment and retention in preparing your fire department for present and future success.
- Develop a comprehensive understanding of your fire department's effectiveness in the recruitment and retention of members.
- Develop an understanding of the specific challenges that your fire department faces in the recruitment and retention of members.
- Determine the reasons why individuals are attracted to join your fire department and their resulting expectations.
- Determine the reasons why members retain their membership in your fire department or decide to leave.
- Evaluate and, as appropriate, utilize new informational media in attracting new members.
- Develop and implement a proactive approach to personnel recruitment and retention.

Things NOT to Do:

- Fail to prepare job descriptions and job specifications for all positions within your fire department.
- Continue to seek candidates only from traditional recruitment sources.
- Address recruitment and retention issues in a reactive, rather than proactive, manner.

Recruiting and Retaining Fire Department Personnel (Job Aid 11)

1. Identify the problems that your fire department has experienced in the recruitment of personnel.

2. Identify the problems that your fire department has experienced in the retention of personnel.

3. Identify strategies that your fire department could use to increase the effectiveness of its personnel recruitment.

4. Identify strategies that your fire department could use to increase the effectiveness of its personnel retention.

Lesson 12: Motivating and Empowering Personnel

Role in Survival and Success

Motivation and empowerment are two essential elements of the effective recruitment and retention of fire department personnel. Likewise, the successful motivation and empowerment of fire department members will contribute to their job satisfaction and job performance. It is, therefore, imperative that you manage and lead your fire department in a manner that creates a positive organizational climate wherein fire department members are motivated and empowered to perform their respective jobs in accordance with the needs of the fire department and the community that it serves.

As fire chief, you will find that motivation and empowerment are related activities that have the potential of greatly enhancing the performance of the individual members of your department, as well as that of the overall organization. Your success in motivating and empowering fire department members will be instrumental in enhancing your department's corresponding success in terms of personnel recruitment and retention. Given that the work of your fire department is in reality performed by individuals working in groups or teams on behalf of the organization, the importance of motivating and empowering personnel in determining your success as the fire chief and that of your department should be obvious.

What You Need to Know

Motivation is a process through which the behavior of an individual—in this case, a member of your fire department—is initiated, directed, and sustained. The importance of motivating department personnel cannot be overstated in that it is through these members that the work of the fire department, both on and off the incident scene, is performed. It is through motivation that you, as the fire chief, will attempt to align the goals and objectives of your members with those of your fire department in the interest of affording members the opportunity to fulfill their own needs while accomplishing work on behalf of your fire department. It is important to begin with an understanding of the motivational needs of the various members of your fire department and the expectations that each member has with respect to the department. A dilemma that you will face in motivating the members of your department results from the fact that whereas motivation is most effective when it is unique and specific to the needs of a given member, it is imperative that you ensure that all members are treated, and perceived as being treated, in a fair and equitable manner.

> ## Motivation
>
> Initiating and sustaining human behavior

Developing an understanding of the expectations of fire department members and the things that serve to motivate them will provide the necessary foundation for effective motivation and will contribute to enhanced success in personnel recruitment and retention. Employ extrinsic and intrinsic rewards as appropriate in the interest of enhancing the job satisfaction, and consequently the job performance, of fire department members. Extrinsic rewards include various types of financial rewards or other incentives, whereas intrinsic rewards derive from the job itself and include task accomplishment, recognition, and status. Intrinsic rewards are typically effective in motivating most fire department personnel, whereas extrinsic rewards are more relevant in motivating personnel serving in a paid capacity.

> ## Types of Rewards
>
> - Extrinsic rewards
> - Intrinsic rewards

Job Performance

The level to which a job incumbent is performing

Job Satisfaction

The satisfaction that a job incumbent derives from his or her job

As fire chief, you and your department will benefit from enhancing your knowledge of motivation through learning more about the various motivation theories and how they can inform your thoughts and actions with respect to motivating your department's personnel. Although a detailed discussion of the various motivation theories is beyond the scope of this book, a summary of the more popular and relevant theories is provided herein. These theories can be classified as content or process theories, in that the content theories consider the factors that motivate individuals in the interest of learning *what* motivates individuals, whereas the process theories focus on the process through which individuals are motivated, thus considering *how* individuals are motivated.

The three primary content theories are the two-factor theory, the hierarchy of needs theory, and the ERG theory. Herzberg's two-factor theory found that factors in the workplace vary in their potential to motivate individuals and can be classified as either hygiene factors or motivating factors. Hygiene factors are those that are necessary to attract an individual to take a job or affiliate with an organization, but do not really serve to motivate the person once in that position. Wages, benefits, and working conditions fall into this category. In contrast, motivating factors are successful in motivating individuals and include such things as status, respect, task autonomy, task identity, and feelings of accomplishment. Hygiene factors are also called dissatisfiers in that although their presence does not play a significant role in motivating members, their absence causes dissatisfaction. Motivating factors are likewise called satisfiers in that their presence contributes to motivation.

Content Motivation Theories

- Two-factor theory
- Hierarchy of needs theory
- ERG theory

Two-Factor Theory (Herzberg)

- Dissatisfiers (hygiene factors)
- Satisfiers (motivating factors)

The hierarchy of needs theory, proposed by Abraham Maslow, classifies the factors that motivate individuals into five sequential levels of need. Maslow advocates that we all start at the lowest level and once we have met or fulfilled that level of need, that level no longer motivates us and we consequently move to the next ascending level as our focus in motivation. The first level in the hierarchy of needs is physiological needs, consisting of such survival necessities as food, water, and shelter. The second level is safety and security needs, which reflects the desire to be safe and secure in one's life and work. The inherent risks associated with fire and emergency services operations makes this level of motivation extremely important and thus one that you as fire chief and your fire officers must continually address. The third level, love and belongingness needs, involves needing to belong, feeling accepted, and enjoying affiliation with others. The climate and culture of the fire department plays a significant role in enabling its members to fulfill this need. Status and self-esteem needs, the fourth level, involve an individual's desire to develop self-respect and the respect of others. Fire department personnel could meet this need level through completing a training course, attaining a professional certification, successfully completing a probationary period, becoming SCBA qualified, becoming an approved apparatus driver/operator, or advancing through the ranks, including becoming an officer. The final need level goes by various names, including self-actualization, self-realization, and self-identification. This level involves an individual achieving his or her full potential.

Hierarchy of Needs Theory (Maslow)

- Physiological needs
- Safety and security needs
- Love and belongingness needs
- Status and self-esteem needs
- Self-actualization needs

Clayton Alderfer's ERG theory is similar to the five-level hierarchy of needs proposed by Maslow, but reduces the number of levels to three: existence needs, relatedness needs, and growth needs. Existence needs encompass the basic needs for survival. Relatedness needs address the need to have interpersonal interactions with others. Growth needs include the need for personal creativity or productive influence.

ERG Theory (Alderfer)

- Existence needs
- Relatedness needs
- Growth needs

Two recognized process theories are expectancy theory and equity theory. Expectancy theory advocates that an individual's behavior is influenced by a number of factors. Individuals consider the effort they would have to exert in order to achieve a given level of performance and the expected outcome they would receive as a consequence. The expectancy of being able to exert the effort, attain the performance level, and achieve the outcome is evaluated by the individual as the basis for determining the value of pursuing a desired outcome.

Process Motivation Theories

- Expectancy theory
- Equity theory

According to equity theory, an individual engages in a mental process through which his or her efforts or inputs and the corresponding or resulting outcomes are weighed or evaluated relative to the efforts and outcomes of other members of an organization. In those cases where the individual feels they are being treated fairly, they will be motivated, whereas should they feel that they are not being treated fairly, their level of motivation and, consequently, their job performance and job satisfaction are likely to decrease.

Empowerment is the act of delegating power and authority to a subordinate in the interest of accomplishing organizational goals and results when a member of your fire department is granted the ability to make decisions within his or her area of responsibility. Through empowerment you enable others within your department to set goals, make decisions, and solve problems. The conditions necessary to support empowerment include motivation, involvement, and participation. Effective delegation, whereby responsibility and accompanying authority is passed from a superior to a subordinate and that subordinate is held accountable to enact the delegated responsibility, provides the foundation for successful empowerment. An additional ingredient of effective empowerment is ensuring that the member receiving the delegated role has access to all the information that is necessary to enact the delegated responsibilities.

Empowerment

Providing organizational personnel the ability to fully utilize their talents and address issues within their area of responsibility

Conditions for Empowerment

- Motivation
- Involvement
- Participation
- Effective delegation
- Information access and availability

Your success as a fire chief and that of your fire department will in large part be determined by your ability to motivate and empower not only those officers who directly report to you, but all fire department personnel. In so doing, you will be creating a positive organizational climate that affirms every member and recognizes and values their contributions to the department. The relationship between recruitment, motivation, empowerment, and retention should be understood. Without motivation, potential members will not apply to join the fire department. Motivation and empowerment are likewise instrumental throughout a member's tenure in the fire department and contribute to job satisfaction and job performance. Motivation and empowerment may also have a significant influence on a member's interest in continuing their affiliation with the fire department. Your understanding of motivation and empowerment and your use of a management and leadership style that contributes to the motivation and empowerment of the members of your department will greatly enhance your effectiveness, as well as that of your fire department.

Job Aid 12 that appears at the end of this lesson is designed to assist you in developing an understanding of what motivates and empowers personnel within your fire department and appropriate strategies for enhancing the motivation and empowerment of your personnel.

Things to DO:

- Recognize the role of motivation and empowerment in the successful recruitment and retention of fire department personnel.
- Recognize the role of motivation and empowerment in contributing to the job satisfaction and job performance of fire department members and consequently to the overall success and performance of your fire department.
- Implement motivational strategies in a manner that is fair and impartial, while remaining responsive to the needs of individual fire department members.
- Use intrinsic and extrinsic rewards as appropriate.
- Recognize the conditions necessary to support the empowerment of fire department members.
- Utilize a leadership approach that creates a positive organizational climate, resulting in the motivation and empowerment of fire department members.

Things NOT to Do:

- Fail to assist fire department members in aligning their needs with organizational goals.
- Fail to understand the expectations of fire department members.

Motivating and Empowering Fire Department Personnel (Job Aid 12)

1. Identify the challenges that your fire department has faced in motivating its personnel.

2. Identify the challenges that your fire department has faced in empowering its personnel.

3. Identify strategies that your fire department can use to enhance its effectiveness in motivating its members.

4. Identify strategies that your fire department can use to enhance its effectiveness in empowering its members.

Lesson 13: Managing Role-Related Issues

Role in Survival and Success

As the fire chief, you will rely on the members of your fire department, working in groups or teams, to perform the work of the department both on and off the incident scene. The success of your fire department in achieving its mission and goals in an effective, efficient, and safe manner while meeting and, where possible, exceeding realistic stakeholder expectations will require that each member of your department correctly understand and enact his or her appropriate roles and responsibilities. The failure to properly define roles and communicate these roles to the members of your department can undermine its success, as well as yours as fire chief.

It is thus imperative that all jobs be properly defined and staffed with competent, highly qualified personnel that fully understand the roles and responsibilities associated with their respective positions and successfully enact those roles and responsibilities. As fire chief, it is important that you recognize the potential for role-related issues that can potentially compromise your organization's success and ensure that all roles are clearly defined, communicated, and enacted.

What You Need to Know

The successful enactment of an interrelated set of roles by fire department members is essential to the effective, efficient, and safe operation of your fire department and to enabling the organization to realize its mission and goals, including meeting and, where possible, exceeding realistic stakeholder expectations. It is thus imperative that you ensure that the roles and responsibilities of all positions within your fire department are clearly defined, thoroughly communicated, and properly and fully enacted. This applies to the roles and responsibilities of fire department members both on and off the incident scene.

Unfortunately, it is possible for role-related issues to arise within your fire department, which can create confusion and conflict, as well as compromise the effectiveness, efficiency, and safety of departmental operations. These issues originate in the delineation, communication, acceptance, and enactment of roles within the organization. The role that defines and delineates the expectations of the fire department with respect to a particular position is called the *sent role*. This role should be based upon the scope of the operations of your fire department and should be reduced to writing in the form of a job description that outlines the duties, responsibilities, activities, and working relationships for this position (figure 13–1).

Role Theory

- Sent role
- Received role
- Enacted role

Sent Role

The role intended by an organization for a particular position

TITLE: CHIEF FIRE OFFICER/FIRE MARSHAL

SUMMARY:

This is highly responsible managerial work involving the overall administration of the Fire Department that consists of volunteer fire companies and a small central staff.

Work involves responsibility for the overall operation of the Fire Department, including the coordination of Fire Department activities with other fire departments, public safety agencies, and Township departments. Duties include the response to all major incidents on a twenty-four hour basis. This position has the responsibility for making difficult field decisions as well as administrative decisions. The work requires that the employee have thorough knowledge, skill and ability in every phase of the fire service management.

SUPERVISION RECEIVED:

Works under the general direction of the Township Manager.

ESSENTIAL FUNCTIONS:

Maintains liaison with and establishes control of several volunteer fire companies that provide the fire suppression services for the Township. Meets regularly with volunteer company Fire Chiefs and Presidents to establish goals and objectives for the Fire Department. Plans strategies for achieving goals. Establishes operating policies and procedures and coordinates implementation in volunteer companies.

Establishes standards and monitors operations to ensure compliance. Conducts inspections of Fire Company facilities, operations, equipment and personnel.

Supervises fire suppression operations at all major incidents. Ensures proper functioning, staffing and equipment. Ensures safe operations.

Directs the activities of subordinate staff and other supervisors. Holds staff meetings. Plans and coordinates departmental activities.

Performs the duties of Fire Marshal. Supervises fire prevention operations, including code enforcement, investigations, plan review and code complaints. Processes requests for variances from fire codes. Manages the fire alarm and fire drill programs. Supervises the investigation of all fires; gives depositions, and provides expert testimony in court. Makes recommendations on fire code changes.

Supervises the investigation of all hazardous materials incidents. Oversees required record keeping and reporting pertaining to hazardous materials.

Directs the preparation and implementation of the Fire Department budget. Presents budget proposals to Township Manager.

Figure 13–1. Job description. (Courtesy of Lower Merion Fire Department)

Attends meetings with other department heads, public officials and community leaders. Sits as a member of, or chairs, various task forces and committees.

Maintains liaison with civic, business, community and professional organizations. Attends meetings. Makes presentations on fire prevention, fire safety and fire suppression.

Oversees training program development and implementation.

Responds to and takes command of multiple-alarm fires, fatal fires or other unusual occurrences.

Performs related work as required.

QUALIFICATIONS:

A bachelor's degree in fire administration or a related field, plus ten years of progressively responsible fire services experience including at least five years in a command position. Graduation from the Executive Fire Officer's course of the National Fire Academy is preferred.

Valid automotive driver's license.

Thorough knowledge of fire service administration, including personnel management, fiscal affairs, program planning, and training.

Thorough knowledge of fire suppression and fire prevention theories, principles, techniques, and equipment, including national, state and local fire codes.

Considerable knowledge of planning and policy research.

Considerable knowledge of hazardous materials control.

Considerable ability to direct the operations of a large, almost completely volunteer fire service under routine and emergency circumstances.

Considerable ability to make decisions, often under pressure or crisis conditions, that may involve life, property or the operations of the entire department.

Considerable ability to identify, select, train, and delegate work to volunteer Fire Chiefs, and central staff.

Considerable ability to express ideas clearly and concisely, both verbally and in writing.

Considerable ability to establish and maintain effective working relationships with superiors, associates, subordinates, volunteer fire companies, officials of other agencies, and the general public.

Figure 13–1. Job description. (Continued)

PHYSICAL REQUIREMENTS:

Ability to sit for a minimum of 4 hours, walk for a minimum of 2 hours, stand for a minimum of 1 hour and drive for a minimum of 1 hour.

Ability to bend, stoop, squat, crawl, climb, crouch, kneel and lift up to 35% of the workday.

Ability to reach above shoulder level, push and pull up to 35% of the workday.

Ability to lift and carry up to 10 pounds of tools, equipment and materials 100% of the workday.

Ability to lift and carry up to 34 pounds of tools, equipment and materials 35% of the workday.

Ability to lift up to 74 pounds of tools, equipment and materials 35% of the workday.

Ability to use both the right and left foot to operate foot controls of vehicles.

Ability to use both the right and left hand for firm grasping and repetitive actions.

Ability to work on unprotected heights and work around moving machinery.

Ability to drive automotive equipment.

Ability to withstand exposure to dust, fumes, gases and noise.

Ability to perform job functions in adverse weather conditions at times.

Figure 13–1. Job description. (Continued)

The job description should serve as the basis for developing a job specification, which delineates the necessary qualifications to succeed in the position defined in the job description. The job specification is, therefore, a valuable tool in recruiting and selecting a candidate to fill the position, whereas the job description serves to inform the successful candidate what the organization expects of him or her in performing that job. The resulting understanding of his or her role on the part of the job incumbent is referred to as the received role. The role that the job incumbent actually performs or enacts is called the enacted role.

Received Role

The understanding a job incumbent has with respect to his or her role

Enacted Role

The role that a job incumbent actually performs

Although the roles and responsibilities of a particular position or assignment often appear obvious, it is essential that you ensure that all such assignments are made in a manner wherein the person accepting the responsibilities and all other relevant and interested individuals, both within and outside your fire department, have an accurate understanding of that person's role within the fire department. The appointment of a departmental training officer or safety officer would serve to illustrate this point. There will be one or more times during your service as fire chief when you will appoint highly qualified individuals to serve in both of these mission-critical positions.

It would seem that the responsibilities of a department training officer are fairly straightforward—however, this may not be the case. The training officer is obviously responsible for "training," but several questions would illustrate the range of related but diverse responsibilities that could be assigned to that position. Is the training officer responsible for scheduling training within the department that is conducted by departmental personnel; for arranging for outside instructors to conduct training within the department; for conducting training within the fire department; for approving departmental members to attend outside training courses and programs; for approving members attending conferences and seminars; for professional development of fire department personnel? Does the training officer have a training budget?

All of these questions, and likely several more, serve to illustrate the importance of properly designing jobs, selecting individuals with the requisite knowledge, skills, and attitudes to staff these positions, providing job incumbents with an accurate and thorough understanding of the roles and responsibilities of their position, and communicating this information to all appropriate individuals, groups, and organizations in the interest of avoiding role-related problems that can undermine the effectiveness of the job incumbent, as well as lead to frustration on his or her part and potentially diminished job performance and job satisfaction.

An additional example would be the appointment of a fire department safety officer. Obviously, this is a crucial position within any contemporary fire department and an appropriate individual must be selected to enact an array of mission-critical roles. Appointing an individual who is both qualified and interested in serving in this role is of paramount importance to the successful execution of the responsibilities of this position within your fire department. Relevant NFPA standards with respect to fire department health and safety and the roles and responsibilities of the safety officer provide invaluable guidance in structuring the roles and responsibilities of this position and selecting a highly qualified and motivated candidate to staff it. Once again, the roles and responsibilities must be clearly defined and understood by all relevant individuals, groups, and organizations. As with the appointment of a training officer, address and clarify a number of issues prior to recruiting, selecting, and appointing an individual to serve in this position. What responsibilities will this position have on the incident scene? What responsibilities will this position have off the incident scene? What role does this position have with respect to the department's infection control program? What authority does this position have within the fire department? Although it is important to recognize that there are health and safety issues in addition to the provision of an incident safety officer at emergency incidents, including infection control, that all fire departments must address, your department may find it prudent to bundle or combine these responsibilities under one position or assign them to separate positions. In either case, devoting the time and effort to properly structure and communicate the roles and responsibilities of the associated position(s) will stand you as fire chief and your fire department in good stead.

Two types of role-related issues can potentially arise within an organization. They are referred to as role ambiguity and role conflict. Role ambiguity occurs when there is a discrepancy between the understanding of the organization and the position incumbent with respect to the role of the position. An example of this is a new firefighter not fully understanding his or her station duties. When the position incumbent fails to act in accordance with his or her understanding of the role, role conflict exists. The new firefighter refusing to perform the expected station duties is an example of role conflict.

Role-Related Issues

- Role ambiguity
- Role conflict

Role Ambiguity

A role-related issue involving a discrepancy between the sent role and the received role

Role Conflict

A role-related issue involving a discrepancy between the received role and the enacted role

An additional consideration with respect to anticipating and avoiding role-related issues is the matter of the "inside" and "outside" roles of the fire chief and other positions within your fire department. This issue, which was discussed previously in Lesson 3, must be understood and addressed in the interest of avoiding the potential role-related issues of role ambiguity and role conflict. Both internal and external stakeholders will have expectations with respect to the roles of the various positions within your fire department. Fire department members may not fully understand and appreciate the necessary "outside" roles of the fire chief within the community or larger fire service. External stakeholders may likewise have expectations with respect to the roles and responsibilities of the fire chief outside the fire department. It is thus imperative to your success and survival as a fire chief, and to that of your fire department, that you determine and enact an appropriate balance between your inside and outside roles, as well as the similar roles of other positions within your fire department.

Balancing Organizational Roles

- "Inside" roles
- "Outside" roles

The importance of understanding and clarifying roles within your fire department cannot be overemphasized. As fire chief, you must ensure that all positions within your fire department are properly defined through the use of job descriptions and that appropriate candidates are selected to staff each position in accordance with the qualifications delineated in an accompanying job specification. The use of effective communication and providing "realistic job previews" will minimize role-related issues resulting from role ambiguity. Proper employee selection and effective supervision will prove essential in preventing and resolving issues arising from role conflict (figure 13–2).

Figure 13–2. Fire chief clarifying job responsibilities with subordinate

Job Aid 13 that appears at the end of this lesson is designed to help you anticipate and address potential role-related issues within your fire department.

Things to DO:

- Recognize the importance of properly defining the roles of all fire department personnel.
- Recognize the potential for role-related issues that can compromise the success of your fire department.
- Develop job descriptions and accompanying job specifications for all positions within your fire department.
- Staff all positions within your fire department with highly qualified personnel based on the qualifications articulated in the appropriate job specifications.
- Ensure that all members of your fire department fully understand their roles and responsibilities.

Things NOT to Do:

- Fail to properly define and communicate the roles and responsibilities of all fire department positions to ensure that all members fully understand their roles and responsibilities.
- Engage in staffing decisions that result in the filling of positions with unqualified individuals.
- Allow role ambiguity and role conflict to arise as a consequence of ineffective communication, selection, or supervision.

Managing Role-Related Issues within the Fire Department (Job Aid 13)

1. Identify role-related issues involving role ambiguity that exist in your fire department.

2. Identify role-related issues involving role conflict that exist in your fire department.

3. Identify strategies that your fire department could use to address issues of role ambiguity.

4. Identify strategies that your fire department could use to address issues of role conflict.

Lesson 14: Managing Performance

Role in Survival and Success

The performance of department members, although an integral determinant of organizational effectiveness and efficiency, is often not properly addressed within contemporary fire departments. Given the essential role that individuals play in enacting the roles and responsibilities necessary to enable the fire department to succeed, the importance of evaluating or appraising their performance should be obvious, but often is overlooked or minimized.

Whether your department is a volunteer, career, or combination department, its success depends on the individual and collective efforts of its members. It is thus important that the performance of job incumbents be evaluated periodically. Regardless of whether a fire department member is serving in a volunteer or a paid capacity, the roles and responsibilities of his or her position should be delineated in a written job description that serves as the basis for subsequent performance evaluation. Your commitment as fire chief to institute an ongoing system of performance evaluation will be instrumental in preparing and positioning your fire department for present and future success.

What You Need to Know

The management of performance within your fire department is a key issue in terms of ensuring the effectiveness, efficiency, and safety of your department and its operations. The success of your fire department, both on and off the incident scene, will be determined by the contributions of each and every member as they enact their specified roles and responsibilities. Together their efforts create a synergy that contributes to the fire department's success. A failure of any member to fully enact his or her roles and responsibilities can compromise the department's effectiveness, efficiency, and safety. It is, therefore, imperative that each member be given clearly defined responsibilities and be held accountable for their performance.

All fire department members, whether volunteer or paid, should be provided with a job description and should be periodically evaluated in accordance with it. Although many fire departments fail to consider the use of performance appraisal for members serving in volunteer capacities, to not do so can be detrimental to the fire department and its ability to effectively, efficiently, and safely serve the community. Several examples serve to clearly illustrate this point.

Who Should be Evaluated?

- Personnel serving in volunteer capacities
- Personnel serving in paid capacities

An administrative member of the fire department could be assigned to research potential grant opportunities to secure funding for the needed replacement of the department's self-contained breathing apparatus (SCBA) and to prepare and submit a grant proposal to seek such essential funding. If that individual failed to complete the assignment or to prepare and submit the grant application according to the submission requirements of the particular grant program, such as the Assistance to Firefighter's Grant Program, this would represent a significant opportunity loss to the fire department in that the successful enactment of this responsibility could have enhanced the department's personnel safety, as well as its operational effectiveness and efficiency.

The firefighters within the department have a number of important responsibilities, including ensuring that all apparatus and equipment is properly restored to service following its use and maintained in a state of readiness. Whether the firefighters are serving in paid or volunteer capacities, this is a mission-critical responsibility that needs to be assigned and for which members need to be held accountable through performance appraisal. This is particularly true in the case of ensuring the care and readiness of personal protective equipment, including self-contained breathing apparatus. Whereas periodic performance evaluations should take place on a scheduled basis, critical performance issues that relate to operations and safety should be addressed as and when necessary.

An additional example of the need for performance evaluation, regardless of a member's status of paid or volunteer, falls in the realm of training. As fire chief you appoint one of your fire officers to be responsible for the training in your department and to function as the departmental training officer. As with any other position, you periodically evaluate this individual's job performance. An essential element of your responsibility to your fire department, its personnel, and the member accepting the crucial responsibility for training your personnel is to take the time to conscientiously evaluate that person's job performance—whether good or bad. In the case where there are deficiencies in performance, discuss them and set clear expectations, but also demonstrate your support for the job incumbent (figure 14-1). It is equally important to conduct a performance evaluation when the job incumbent is meeting or exceeding the performance expectations for the position, in the interest of recognizing and affirming this performance and consequently contributing to the member's motivation and empowerment.

When is Performance Evaluation Important?

- When there are deficiencies in performance
- When performance is satisfactory
- When performance exceeds expectations

Figure 14–1. Fire chief conducting performance appraisal. (Courtesy of Carole Taylor)

The roles and responsibilities of a given position, along with the ways that position contributes to the overall operations and success of the fire department, should be clarified and thoroughly understood by the job incumbent at the time he or she is offered the position. It may also be necessary to review these roles, responsibilities, and performance expectations at the beginning of a performance appraisal session. This should be followed with a discussion regarding the job incumbent's actual performance in light of the department's performance expectations for that position. In those instances where the individual's performance is meeting expectations and performance is thus satisfactory, this should be acknowledged and commended. When the member's performance is exceeding the department's expectations, this should be affirmed and should open the door to discussion of the individual's future interests within the department. When performance deficiencies are identified, it is important that the member understand what is expected, where performance is falling short, what needs to be corrected in terms of performance, and the consequences if the desired performance level is not achieved.

Performance Appraisal Session Format

- Review performance expectations.
- Review actual performance.
- Compare performance to expectations.
- Recognize successful performance.
- Identify performance deficiencies.
- Identify areas for improvement.
- Discuss consequences of continued performance deficiencies.
- Discuss strategies for improving performance.

Through such a periodic evaluation process, potential problems, including job performance issues, can be identified and addressed. The additional problems or issues that can be identified and addressed through performance evaluation include: attendance, tardiness, misconduct, and training inadequacies. The use of performance evaluations also provides the feedback and encouragement needed by fire department members to engage in self-development. The use of performance evaluation, or performance appraisal as it is also called, thus represents a proactive approach to the identification and resolution of job performance issues.

Typical Problems Addressed in Performance Evaluations

- Job performance
- Attendance
- Tardiness
- Misconduct
- Training needs

The criteria and all performance measures utilized in conducting a performance evaluation should be job-related and clearly specified in writing. The job description for a particular position should serve as the basis for the performance evaluation of the individual serving in that position. The performance evaluation process should be conducted in a fair and objective manner that builds trust between the supervisor and the member.

Criteria for Conducting Performance Appraisals

- Fair
- Objective
- Consistent
- Professional
- Based on job-specific criteria

Logistics are extremely important and dictate that you select an appropriate location and time for the performance review session. The venue for a performance appraisal session should provide for a one-on-one private discussion between the fire department member and his or her supervisor. It should also be scheduled for a time that is mutually convenient and ensures that necessary and sufficient time is available to have a meaningful discussion regarding the individual's performance, which, as appropriate, addresses both the positives and negatives of that performance. Necessary arrangements should be made to avoid interruptions during the performance appraisal session. The inclination to substitute informal discussions regarding a member's performance, which are conducted in front of others in public areas such as the engine bays of the firehouse, should be resisted in that such an approach can be interpreted as minimizing the importance of job performance as well as sending the wrong message or impression regarding a member's value to the fire department. Taking time to properly prepare for and conduct a formal periodic performance evaluation is thus beneficial to both the job incumbent and your fire department.

Although a number of areas for improvement may be identified and discussed during the performance evaluation interview, it is important to affirm the member in terms of areas in which his or her performance meets or exceeds the department's expectations. It is also important to provide and encourage ample opportunity for the member to express any questions they may have regarding the roles, responsibilities, and expectations of their job, as well as to communicate any obstacles they have experienced that may have prevented them from successfully performing their job and meeting departmental expectations. Additional support, including necessary training or resources, that would enable the job incumbent to succeed should also be discussed during the performance appraisal interview.

It is important when conducting a performance appraisal to thoroughly prepare in advance of the counseling session and to conduct yourself professionally throughout the session. The focus of the discussion should be on performance issues rather than personalities, with a positive approach being utilized throughout the counseling session. You should conclude the performance appraisal by summarizing the member's performance, establishing areas for improvement, and discussing any development needs. You will find that a proper counseling attitude, supported by honesty and integrity, will contribute to a favorable outcome in performance appraisals in terms of affirming those members whose performance meets or exceeds organizational expectations and motivating those with performance deficiencies to improve their performance.

Focus of Performance Review Session

- Focus on performance issues
- Do not focus on personalities

Job Aid 14 that appears at the end of this lesson is designed to assist you in considering how the use of performance evaluation could be improved within your fire department

Things to DO:

- Recognize the importance of the periodic evaluation of the performance of all members of your fire department.
- Establish and implement a performance evaluation process within your fire department.
- Periodically evaluate the performance of all personnel within your fire department.
- Use job-related and specific measures when evaluating member performance.
- Focus on performance rather than personalities when conducting performance interviews.
- Conduct performance evaluations in a fair and consistent manner.
- Conduct yourself professionally during performance evaluation interviews.
- Summarize performance, deficiencies, and planned actions when concluding a performance evaluation session.

Things NOT to Do:

- Fail to recognize the value of performance evaluation to the present and future success of your fire department.

- Assume that there is no need to evaluate personnel serving in a voluntary capacity.

- Fail to prepare prior to conducting a performance evaluation interview.

- Fail to consider proper logistics when conducting performance evaluation sessions.

Evaluating Performance of Fire Department Personnel (Job Aid 14)

1. Describe the current use of performance evaluation in your fire department.

2. Discuss how the use of performance evaluation could be improved within your fire department.

3. Discuss the benefits of changing your fire department's approach to performance evaluation.

4. Describe the strategies that you, as fire chief, would use to enhance the use of performance evaluation within your fire department.

Lesson 15: Administering Discipline

Role in Survival and Success

One of the most undesirable and often challenging issues that you will have to deal with as a fire chief is the administration of discipline to members of your fire department. Although it is rare that fire chiefs look forward to handling such matters, the timely and effective resolution of disciplinary matters is crucial to your success as a fire chief and also to that of your department.

Discipline is intended to ensure that the behavior of all members complies with the rules, regulations, policies, and procedures of your fire department. Each of these administrative documents articulates the expectations of your fire department with respect to the behavior of its members. Compliance with these behavioral expectations ensures that fire department members are working towards the common good of achieving the mission and goals of the fire department and consequently effectively, efficiently, and safely meeting and, where possible, exceeding the realistic expectations of departmental stakeholders.

A properly articulated and enforced system of discipline is thus essential to the present and future success of your fire department. The absence of appropriate discipline can conversely undermine your fire department's success.

What You Need to Know

The successful operation of any contemporary organization, including your fire department, requires that in addition to each and every member of the organization fully understanding and enacting their defined roles and responsibilities, they do so in compliance with all rules, regulations, policies, and procedures developed and promulgated by the organization. The information provided in these various administrative documents is designed to clarify your fire department's expectations with respect to the behavior of its members. Members routinely engage in compliant behavior, but there may be those individuals and instances where behavioral expectations are violated. In such situations, timely action is required in the interest of correcting the inappropriate or unacceptable behavior of fire department members, while reinforcing the importance of compliance on the part of other department members.

> **Administrative Documents that Articulate Behavioral Expectations**
>
> - Policies
> - Procedures
> - Regulations
> - Rules

Your intent in administering discipline will normally be to encourage the self-correction of unacceptable behavior on the part of fire department members. This is the case in most situations, the exception being serious or critical infractions that rise to the level wherein the member should be terminated from the organization. These types of violations or infractions should be clearly delineated in the administrative documents of your fire department.

> **Discipline**
>
> Process of establishing, communicating, and enforcing behavioral guidelines

An approach to the administration of discipline that is utilized by many contemporary organizations, including fire departments, is called progressive discipline. This disciplinary approach is designed to call attention to unacceptable or inappropriate behavior and motivate the involved fire department member to engage in self-correction of that behavior. This approach to discipline is called *progressive* in that the penalties associated with continued infractions increase to the eventual termination from the fire department. It should be understood that the intent of using progressive discipline is to use increasing penalties to get the member's attention and to motivate him or her to correct the problem behavior rather than to "get the goods on the member" to justify terminating him or her from the department.

An example of the use of progressive discipline is a fire department member who exhibited a problem with lateness. Through progressive discipline, a series of disciplinary actions that increase in severity should the undesirable behavior continue, can be used to get the member's attention in order to reinforce the importance of correcting this behavior and to encourage the member to take the necessary steps to correct this behavior.

Progressive Discipline

Process through which the consequences of disciplinary action increase in the event that undesirable behavior is not corrected

The use of such an approach to discipline should be based on the prior establishment of a progressive discipline process, which should be communicated to and understood by all members of the fire department. Likewise, these members should have a full understanding of the department's rules, regulations, policies, and procedures, as well as the corresponding behavioral expectations for members. A typical progressive discipline procedure might consist of four steps: (1) a verbal warning, (2) a written warning, (3) one or more suspensions, and (4) termination. The intent in using such a procedure should be to encourage behavior correction at the lowest possible level through self-correction on the part of the member. Other progressive discipline models incorporate transfer or demotion, alternatives which may not be practical or desirable in many fire departments.

Typical Progressive Discipline Procedure

1. Verbal warning
2. Written warning
3. Suspension(s)
4. Termination

Although you will find that progressive discipline is a valuable approach to the administration of discipline, there will be those instances where members do not correct their behavior and eventually reach the step in the progressive discipline procedure where their termination from the organization is appropriate. There are also critical offenses that may rightfully result in immediate discharge rather than the utilization of the progressive discipline penalties. It is important that such infractions be clearly articulated in your department's progressive discipline policies and procedures.

Your skills in handling the administration of discipline will develop over time, as will your confidence in addressing such matters. Discipline will always represent an unenviable task, and with time you will discover that you are becoming more capable and confident in this dimension of your responsibilities as fire chief. You will also realize the importance of the proper administration of discipline in motivating, empowering, and retaining fire department personnel.

Job Aid 15 that appears at the end of this lesson is designed to assist you in considering how discipline is administered within your fire department.

Things to DO:

- Recognize that although administering discipline may present certain inherent challenges, it is essential to the present and future success of your fire department.

- Ensure that all members of your fire department are fully aware of the department's rules, regulations, policies, and procedures, including those that relate to member discipline.

- Approach the administration of discipline with a mindset of assisting the member in attaining compliance with departmental behavioral expectations rather than the necessary process to follow to terminate a member.

- Develop and implement a progressive discipline process and procedure that meets the needs of your department in terms of encouraging desired behavior and compliance with behavioral expectations.

- Avoid the tendency to not address disciplinary issues.

- Administer discipline in a fair and unbiased manner, consistent with departmental policies and procedures.

- Use appropriate discipline as a tool to enhance the motivation and empowerment of fire department members as well as the organizational climate of your fire department.

Things NOT to Do:

- Fail to ensure that all fire department members fully understand the behavioral expectations of the department and its disciplinary policies and procedures.
- Fail to address all disciplinary issues in a timely and professional manner in accordance with departmental policies and procedures.
- Refrain from terminating members for critical offenses or exhaustion of the remedies within your department's progressive discipline procedure.

Administering Discipline within the Fire Department (Job Aid 15)

1. Describe your fire department's current approach to administering discipline.

2. Identify all relevant administrative documents that address the issue of administering discipline within your fire department.

3. Discuss whether your fire department has a progressive approach to discipline. If so, identify the steps in your progressive discipline procedure.

4. Discuss how you would improve your fire department's approach to the administration of discipline.

Lesson 16: Communicating Effectively

Role in Survival and Success

A significant amount of your work as fire chief will involve communicating with individuals and groups within and outside your fire department. Your success in managing and leading your fire department and enacting each of the associated management functions require you to have effective communication skills. As the fire chief, you are expected to enact an array of roles, most of which incorporate the use of communication.

Communication will, therefore, be instrumental in determining your success as fire chief, as well as that of your fire department. Your understanding of the principles and practices of effective communication and skill in their application will be of paramount importance as you interact with fire department members and the other stakeholders of your fire department. The use of effective communication will position your organization for success, whereas the inability to communicate effectively will present significant challenges to your professional success as fire chief and to the success of the fire department that you have been afforded the opportunity to manage and lead.

What You Need to Know

Communication is the process through which individuals attempt to share meaning through the transmission of messages. The goal of communication is to develop a shared understanding on the part of the individual initiating the message and the individual who receives it. Communication is thus considered to be effective when it results in a shared understanding between the parties to the communication process, and likewise ineffective when the desired shared understanding does not result.

Communication

Process of exchanging thoughts and ideas between a sender and a receiver in the interest of achieving a shared understanding

Goal of Communication

Develop a shared understanding between the initiator and the recipient of a message.

Your success as a fire chief will be enhanced through your understanding of the communication process and skill in communicating through the use of this process. The communication process involves two parties: the sender, who possesses an idea or thought that he or she desires to share, and the receiver, who is the recipient of that communication. Through an encoding process, the sender converts that idea or thought into a message, which is sent through an appropriate message channel. The receiver utilizes a decoding process to form an understanding of the message. If the process results in a shared understanding between the two parties, effective communication has resulted. The effectiveness of the communication process can be further enhanced through the use of feedback between the receiver and the sender, resulting in two-way communication between the parties (figure 16–1).

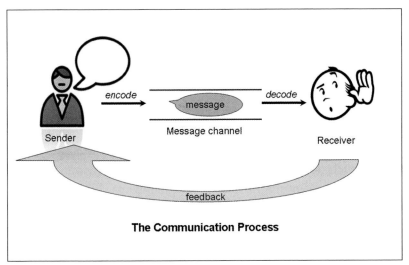

Figure 16–1. Communication process

Types of Communication

- One-way communication
- Two-way communication

The communication that occurs within your fire department can be categorized as formal or informal, with formal communication being initiated by the fire department as an organization and informal communication being initiated by members of the department. Formal communication is thus official, whereas informal communication is unofficial.

Types of Communication

- Formal communication
- Informal communication

Your roles and responsibilities as a fire chief will routinely require you to engage in formal communication through the issuance of written documents such as policies, procedures, memorandums, and letters. The policies you develop and implement will provide broad or general guidance and serve as a guide for decision making, whereas the procedures you issue will provide specific direction as to how to perform particular activities or evolutions, thus serving as a guide for action. As an example, your department could have a policy that articulates its general position on safety and supporting procedures relative to personal accountability and the utilization of an incident safety officer.

Formal Communication

- Policies
- Procedures
- Orders
- Directives

Policy

Guide for decision making that is broad and general in content

Procedure

Guide for action that is specific
in content

The formal communication that you initiate as fire chief will also include orders and directives. As with policies and procedures, you need to understand the difference between orders and directives. You issue orders on the incident scene as you direct personnel and units to perform the tasks necessary to successfully resolve an emergency situation. Directives are typically issued off the incident scene and relate to assignments and tasks such as station duties or public education activities.

Order

Direction given at an incident scene from
a superior to a subordinate

Directive

Direction provided by a supervisor to a
subordinate off the incident scene

Communication within a fire department that arises outside of formal channels is referred to as informal communication and includes such things as rumors, gossip, and "the grapevine." You will come to appreciate the potential that this type of communication has to negatively impact a fire department's internal climate, morale, and effectiveness. The key to minimizing the potential negative effects of informal communication is to adopt a management approach wherein accurate official information is always provided in a timely manner, thus minimizing or eliminating the need for members to buy into or believe rumors and other inaccurate forms of informal communication.

The two primary forms of communication that you will utilize as a fire chief are written communication and verbal communication. You use both forms of communication as you communicate with those within and outside your fire department. Written communication will include memorandums designed to communicate with departmental members and letters directed to individuals or entities outside your fire department, as well as various reports that you will prepare throughout your tenure as fire chief. The professionalism of your communication, whether written or verbal, is a reflection of you as fire chief and of your department. It is, therefore, beneficial when preparing important written correspondence and documents to have a second person proofread your work for accuracy, readability, spelling, grammar, and punctuation.

Forms of Communication

- Written communication
- Verbal communication
- Nonverbal communication

Much of your communication as a fire chief will be verbal in nature, involving face-to-face conversations or those mediated through the telephone or more advanced technologies. Verbal communication is usually accompanied by nonverbal communication, such as body language, voice characteristics, or appearance. It is important to realize that your nonverbal communication will either affirm and support your words or refute and contradict them. Your professional success as a fire chief will demand that you become an effective communicator capable of successfully utilizing the verbal, nonverbal, and written communication avenues available to you.

Your communication effectiveness can be enhanced through the development of an understanding of the potential barriers to effective communication and the utilization of appropriate strategies and techniques to successfully address these barriers. The potential barriers to effective communication include attributes of the sender, attributes of the receiver, the interaction between the sender and the receiver, and environmental factors.

Barriers to Effective Communication

- Attributes of the sender
- Attributes of the receiver
- Sender/receiver interaction
- Environmental factors

Attributes of the sender that can contribute to ineffective communication include the use of technical words or jargon, confusing or inconsistent messages, and a lack of credibility. Poor listening skills, selective perception, and predispositions comprise the attributes of the receiver that may compromise communication effectiveness. Semantics problems and status differences result in communication ineffectiveness based on the interaction between the parties to the communication, whereas factors such as information overload or noise present themselves as environmental barriers to effective communication.

Fortunately, there are proven techniques that you can use as a fire chief to enhance your communication effectiveness. When communicating with others, craft your message so that it is tailored to its specific recipient or audience through paying attention to language and semantics. Maintain credibility and demonstrate empathy as you communicate. Additionally, ensure that your nonverbal communication is consistent with and reinforces that which you communicate verbally.

As the recipient of communication, engage in active listening— listening carefully and deliberately. And be sensitive to the other person's perspective, using two-way communication to enhance the accuracy and effectiveness of the communication.

Your success as a fire chief and that of your fire department will demand that you develop a thorough understanding of the communication process and the inherent challenges involved in communication. Your communication skills will be put to the test throughout your service as fire chief, oftentimes on a daily basis.

Your communication effectiveness will be based on the successful integration of your communication skills and ability, with your experience and expertise in the fire service and in management and leadership. Your roles and responsibilities as fire chief will require you to make presentations to the public and community groups, as well as to elected and appointed officials. There will be times that you will conduct press conferences and be interviewed by the media (figures 16-2 and 16-3). Your writing skills will be utilized in the drafting of letters and memorandum, preparing meeting agendas, and preparing press releases (figure 16-4). The hallmark of your success in communication will be your ability to communicate in a manner that yields a shared understanding on the part of those with whom you communicate, both within and outside your fire department.

Figure 16–2. Fire chief conducting press conference. (Courtesy of Philadelphia Fire Department)

Figure 16–3. Fire chief being interviewed by reporter. (Courtesy of TrafficDan Miller)

PHILADELPHIA
FIRE DEPARTMENT

Media Contact:
Daniel A. Williams
Executive Chief
(O) 215-686-1300
(C) 267-882-7091

Michael A. Nutter
Mayor

Lloyd Ayers
Fire Commissioner

FOR IMMEDIATE RELEASE:

Philadelphia: Fire Commissioner Lloyd Ayers is proud to announce the successful completion of the third year of the Philadelphia Fire Department's (PFD) partnership with the United States Air Force and a new agreement with the University of New Mexico. There will be a ceremony recognizing the completion of the third year of our partnership on June 2, 2010 at 1:30 p.m. at the Fire Administration Building, 240 Spring Garden Street.

Recognizing the opportunity and value of experiencing a high volume urban Emergency Medical Services system, the United States Air Force Pararescue/Paramedic Program and the University of New Mexico have continued the agreement with the Philadelphia Fire Department to continue to be a training site for this elite program.

Graduates of the program become Pararescue Jumpers who are tasked with recovery and medical treatment of personnel in humanitarian and combat environments. They are the only members of the Department of Defense specifically organized, trained and equipped to conduct personnel recovery operations in hostile or denied areas as a primary mission. Pararescue Jumpers are also used to support NASA missions and have been used to recover astronauts after water landings.

Students from the Pararescue/Paramedic program arrived in Philadelphia on April 19, 2010 for orientation and have been assigned to PFD Paramedic preceptors. Prior to their arrival they complete a paramedic education program at the University of New Mexico and several military special operations courses. While in Philadelphia the students attend clinical rotations at several regional hospitals and ride with Fire Service Paramedic preceptors on Fire Department Medic Units.

Joining Commissioner Ayers will be Scott Valenti, Director, USAF Paramedic Program and Jeff Gregor, Director of the University of New Mexico Paramedic Program,

# # #

Figure 16–4. Press release. (Courtesy of Philadelphia Fire Department)

Job Aid 16 that appears at the end of this lesson is designed to assist you in evaluating the effectiveness of communication within your fire department and in identifying strategies for increasing communication effectiveness.

Things to DO:

- Recognize the importance of effective communication in the successful enactment of your roles and responsibilities as fire chief.

- Use timely and accurate formal communication to minimize the potential negative outcomes of informal communication.

- Develop and implement policies and procedures that clearly communicate the intent of your fire department to its members.

- Utilize effective communication techniques in issuing orders and directives.

- Recognize the importance of accuracy and professionalism in all written communications and have a colleague review important written documents before their issuance.

- Ensure that your nonverbal communication is consistent with and supports your verbal communication.

- Develop an understanding of the potential barriers to effective communication and utilize appropriate strategies and techniques to address these issues.

- Tailor your message to its intended recipient or audience.

- Communicate in a credible, honest, and professional manner.

- Use active listening to discern messages that you receive.

- Communicate in a manner that leads to a shared understanding between the parties to the communication.

Things NOT to Do:

- Fail to enhance your communication effectiveness through understanding and applying the communication process.

- Fail to realize that your communication reflects on both you and your fire department.

- Fail to demonstrate honesty, credibility, empathy, and professionalism in all communication.

- Engage in passive, rather than active, listening.

Increasing Communication Effectiveness (Job Aid 16)

1. Identify areas in which your fire department's communication is effective and ineffective.

2. Identify areas in which your communication, as fire chief, is effective and ineffective.

3. Describe strategies that your fire department could use to increase its communication effectiveness.

4. Describe strategies that you could use to increase the personal effectiveness of your communication as fire chief.

Lesson 17: Making Decisions

Role in Survival and Success

The successful management and leadership of your fire department will require you to make frequent decisions, particularly as you enact your responsibilities for planning, organizing, directing, and controlling. Your effectiveness in making decisions will necessitate that you understand the decision-making process and its associated challenges, as well as the merit of making informed decisions based on relevant and useful information and the role that quality information plays in enhancing decision-making effectiveness.

Your roles and responsibilities as fire chief will require you to make decisions that are routine and recurring, as well as those that are infrequent and unique. The utilization of a rational and proven decision-making process will increase your decision-making effectiveness as fire chief, which in turn will contribute to the enhanced success of your fire department. Over time you will come to fully recognize and appreciate the integral role that decision making plays in the management and leadership of your fire department.

What You Need to Know

Decision making involves choosing from available alternatives. It is a process of identifying problems and opportunities and selecting an appropriate alternative to resolve a problem or pursue an opportunity. As fire chief, you will face many situations that require you to engage in decision making that fall under three possible conditions: certainty, risk, and uncertainty. Certainty exists in cases where all of the necessary information to make a decision is available to the decision maker. Under a condition of risk, although information may be available, the outcomes associated with decision alternatives are unknown and thus subject to chance. When needed information is unavailable or incomplete, uncertainty exists.

Decision Making

Process of making informed decisions by selecting from available alternatives

Decision-Making Conditions

- Certainty
- Risk
- Uncertainty

Certainty

When all information necessary to make a decision is available

Risk

When information is available, but outcomes associated with decision-making alternatives are unknown

Uncertainty

When needed information is unavailable or incomplete.

As fire chief, you should strive to make informed decisions based on necessary information that enhances decision-making effectiveness and resulting outcomes. The criteria for "useful" information that supports effective decision making include: relevancy, accuracy, credibility, timeliness, and cost-effectiveness. Relevancy indicates that the information is consistent with decision-making needs. Accuracy relates to the factual correctness of information, whereas credibility considers the source of the information to ensure that it is fair and unbiased. Information used in decision making should be current, as well as available at the time when it is needed within the decision-making process. Lastly, the value of the information in terms of enhancing decision making should outweigh the costs of gathering that information.

Criteria for "Useful" Information

- Relevancy
- Accuracy
- Credibility
- Timeliness
- Cost-effectiveness

As fire chief, your effectiveness in making decisions will be improved by the guidance and structure provided through the use of a proven decision-making process. You will find that the classical model, or rational model as it is also called, will provide that needed decision-making structure through delineating a sequential set of steps designed to contribute to effective decision making. The steps in this model are: (1) recognize the need to make a decision; (2) identify needed information; (3) gather and analyze data; (4) identify alternatives; (5) evaluate alternatives; (6) select an alternative; (7) implement the decision; and (8) monitor and evaluate results. Key points to remember when using this model include the crucial initial step of recognizing the need to make a decision, the importance of actually making and implementing a decision, and the necessity of subsequent evaluation of the results of that decision.

Classical (Rational) Decision-Making Model

1. Recognize need to make decision
2. Identify needed information
3. Gather and analyze information
4. Identify alternatives
5. Evaluate alternatives
6. Select an alternative
7. Implement decision
8. Monitor and evaluate results

In approaching a decision, you will find it prudent to consider the importance and consequences of that decision in the interest of determining if your goal in decision making should be to make the best possible or ideal decision, referred to as maximizing or optimizing, or simply to make a satisfactory and workable decision, known as satisficing. In those situations where you are seeking an optimal decision, it is advisable to fully enact each step within the decision-making process, whereas in those cases where satisficing is acceptable, follow the model only to the point necessary to determine a viable alternative.

Goals in Decision Making

- Maximizing (optimizing)
- Satisficing

Maximizing (Optimizing)

Situation where the goal of decision making is to make the best or ideal decision

Satisficing

Decision-making approach wherein the desire is to make a satisfactory or viable decision, rather than seeking to make the best possible decision

Decisions in your fire department can be made by individuals or groups. The advantages of individual decision making include that it is timely and cost effective. The disadvantages of individual decision making are that important input is often not considered and departmental members who will be required to implement the decision may lack commitment to do so based on their lack of involvement in making it. Group decision making, although more time consuming and costly, can contribute to improved decision making and yield a commitment on the part of involved individuals to the successful implementation of the resulting decision. As fire chief, you will benefit from the conscientious consideration of the appropriateness of individual versus group decision making as you are confronted with decision-making situations.

Decision-Making Involvement

- Individual decision making
- Group decision making

Some of the decisions that will present themselves to you will involve routine and recurring issues. Handle these as programmed decisions, wherein you develop procedures for handling them when they occur, thus freeing up your time and that of the other members of your department to focus attention on the non-programmed decisions, which, based on their uniqueness or infrequency, will often require the thought and contemplation provided through a formal decision-making process.

> **Types of Management Decisions**
> - Programmed decisions
> - Non-programmed decisions

At times it may seem that effective decision makers have the ability to make instantaneous or "snap" decisions when confronted with a decision-making situation. It is important that you not allow yourself to make this false assumption and compromise your effectiveness as a decision maker because it is likely that rather than making a "snap" decision, the seemingly astute decision maker was in reality simply implementing a programmed decision. To fail to recognize this, and not utilize a regimented decision-making process in those situations where it is necessary, will have the potential of greatly compromising the effectiveness of your decision making, and hence your success as fire chief and that of your fire department.

Job Aid 17 that appears at the end of this lesson is designed to facilitate your analysis of decision making within your fire department and to identify ways in which that decision making could be improved.

Things to DO:

- Recognize the important role of decision making in enacting your management and leadership roles and responsibilities as fire chief, including planning, organizing, directing, and controlling.

- Recognize the importance of making informed decisions based on appropriate information and make decisions in this manner.

- Recognize the conditions under which you will make decisions.

- Understand the criteria for "useful" information in making decisions.

- Categorize decisions in terms of their importance and consequence as a basis for either optimizing or satisficing.

- Recognize the advantages and disadvantages of individual and group decision making and utilize an appropriate approach in each decision-making situation.
- Differentiate between programmed decisions and non-programmed decisions, utilizing programmed decisions when and where appropriate.

Things NOT to Do:

- Engage in decision making without the benefit of a logical decision-making process.
- Fail to recognize the need to make a decision.
- Make a decision, but fail to implement it properly.
- Fail to monitor and evaluate the results of decisions subsequent to their implementation.
- Adopt a decision-making approach wherein you attempt to make an optimal or ideal decision in all situations.
- Assume that astute decision makers routinely make "snap" or instantaneous decisions.

Enhancing Decision-Making Effectiveness (Job Aid 17)

1. Identify areas in which your fire department is effective and ineffective in making and implementing decisions.

2. Identify areas in which you, as fire chief, are effective and ineffective in making and implementing decisions.

3. Discuss strategies that your fire department could use to enhance its effectiveness in making and implementing decisions.

4. Discuss strategies that you, as fire chief, could use to enhance your effectiveness in making and implementing decisions.

Lesson 18: Solving Problems

Role in Survival and Success

Throughout your career as fire chief, you will continually be confronted with problems, issues, and challenges that will require you to have sound problem-solving skills. Some of these problems will be fairly inconsequential in terms of their magnitude, scope, and impact; however, others will not in that they may even rise to the level of challenging your success as fire chief and that of your fire department. Problem solving may thus prove to be one of your most demanding and challenging issues as a fire chief.

Develop the ability to ascertain the significance and urgency of the various problems that you and your fire department will encounter, as well as a viable approach to analyzing and resolving these problems. To not do so will itself prove problematic in terms of preparing and positioning your fire department for both present and future success. Problem solving will thus represent one of the most important skill sets that you will want to develop early in your service as fire chief and hone throughout your tenure in this mission-critical position.

What You Need to Know

The problem-solving process follows the decision-making model presented in the previous lesson. Each of the integral steps in the rational decision-making model is likewise valuable in enabling you to effectively and efficiently resolve problems within your fire department. Whereas the first step in this model was earlier identified as recognizing the need to make a decision, when your focus is problem-solving this initial step involves recognizing that a problem exists and framing or developing an understanding of the nature of that problem. As in decision making, this initial step is critical to the success of the overall process and, in this case, to the successful resolution of the involved problem.

Using the Classical Decision-Making Model in Problem Solving

- First step: Recognize that a problem exists and identify that problem.
- Last step: Evaluate whether or not the initial problem has been resolved.

Your success in problem solving will require that you develop a thorough and accurate understanding of the actual problem that faces you. It is fairly easy to confuse its symptoms with the actual problem, and it is thus imperative that you avoid this potential pitfall in your problem-solving activities. The problems that you encounter and address as fire chief on behalf of your fire department may be recurring problems that are generic to departments throughout the fire service or exceptional problems that are unique to your fire department.

Problem

An undesirable situation

Symptom

A situation that may appear to be a problem, but is rather the result of the actual problem

Types of Problems

- Recurring problems
- Exceptional problems

Each of the steps in the previously presented decision-making model will prove instrumental to your success in problem solving, as will their sequential application. The final step when you are utilizing the decision-making model in problem solving is to evaluate whether or not the initial problem has been resolved through the successful implementation of the selected alternative. Three possible outcomes exist: the original problem has been resolved, the original problem has not been resolved, or a new problem has been created. In those situations where either the original problem still exists or a new problem has resulted, it will be important that you revisit the steps of the problem-solving process in the interest of effectively and efficiently resolving the problem as it now presents itself.

Problem-Solving Outcomes

- The original problem has been resolved.
- The original problem has not been resolved.
- A new problem has been created.

Some of the more challenging problems that you will encounter as fire chief may involve the management of conflict and the resolution of complaints. Given their nature, these special areas of problem solving will demand the astute use of your interpersonal skills and should be handled in a timely and professional manner.

Problem-solving situations will often benefit from the successful use of innovation and creativity on your part. This is particularly important when identifying and evaluating possible alternatives. Although some of the problems that your fire department will face will be organization-specific and unique, most will not. The merit of seeking and considering the ideas and input of others as you seek to resolve problems should, therefore, be obvious given the fact that it will be rare that your fire department is facing a problem that is so unique that other departments have not previously faced and attempted to resolve it. In such cases, you stand to benefit from their experience whether or not they were able to successfully resolve the problem.

The informational resources available to you and your fire department through the National Emergency Training Center's

Learning Resource Center will be invaluable in these situations. The use of these resources will enable you to avoid "reinventing the wheel" and consequently enable you to effectively and efficiently address and resolve problems. The research of participants in the National Fire Academy's Executive Fire Officer Program, reported in their Applied Research Projects available through the NETC Learning Resource Center, will provide valuable insights as you conceptualize problems, identify and evaluate possible resolution alternatives, and consider appropriate implementation strategies (figure 18–1).

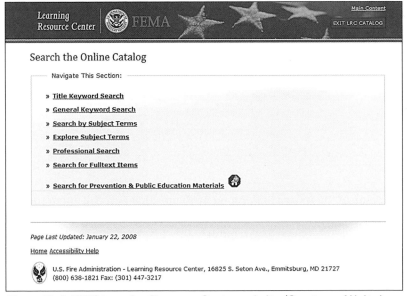

Figure 18–1. NETC Learning Resource Center website. (Courtesy of United States Fire Administration)

Job Aid 18 that appears at the end of this lesson is designed to facilitate your analysis of problem solving within your fire department and identify ways in which your problem solving can be improved.

Things to DO:

- Recognize the importance of problem-solving skills to your success as fire chief and that of your fire department.
- Ascertain the significance and consequence of all problems and act in accordance with this understanding.
- Utilize a proactive approach to organizational problem solving.
- Recognize the possible outcomes of problem solving and appropriately respond to each outcome.
- Recognize the unique challenges of managing complaints and conflict and the importance of interpersonal skills to the successful resolution of these types of problems.
- Resolve all problems in a timely and professional manner.
- Seek and utilize information on how other fire departments have addressed the problems that confront your fire department.
- Utilize the National Emergency Training Center's Learning Resource Center and its collection of Executive Fire Officer Program applied research projects to gain insights on how other fire departments have addressed particular problems.

Things NOT to Do:

- Utilize a reactive, rather than proactive, problem-solving approach.
- Fail to use a comprehensive decision-making process in problem solving.
- Confuse the symptoms of a problem with the actual problem.
- Make the assumption that all of the problems you are facing are unique to your fire department.

Increasing Problem-Solving Effectiveness (Job Aid 18)

1. Identify areas in which your fire department is effective and ineffective in solving problems.

2. Identify areas in which you, as fire chief, are effective and ineffective in solving problems.

3. Identify strategies that your fire department could use to increase its effectiveness in solving problems.

4. Identify strategies that you, as fire chief, could use to increase your effectiveness in solving problems.

Lesson 19: Handling Complaints, Concerns, and Inquiries

Role in Survival and Success

Whereas most fire chiefs would probably categorize handling complaints, concerns, and inquiries as one of the least desirable aspects of their position, it is in reality one of their most important responsibilities. In a world where the focus of most organizations is on customer service, and the organizational stakeholders of your fire department have high expectations in this regard, the timely and professional handling of complaints, concerns, and inquiries, whether originating within the department or from outside sources, is an essential component of your department's responsibilities. With this in mind, customer service should represent a priority within your fire department. Each complaint, concern, or inquiry that your fire department receives represents a customer service opportunity.

The skills required to effectively and efficiently receive and address complaints, concerns, and inquiries are multifaceted and include many of the skills that have been discussed in the previous lessons of this book. Your success, and that of your fire department, in handling these issues will require skillful problem solving coupled with strong interpersonal skills that include the ability to listen attentively and develop a correct understanding of the nature of and contributing factors or reasons that led to the complaint, concern, or inquiry. Your effectiveness in handling these matters will be central to your success and that of your fire department in meeting and, where possible, exceeding the realistic expectations of organizational stakeholders.

What You Need to Know

Responding to complaints, concerns, or inquiries from organizational stakeholders, including fire department members, the public, and elected and appointed officials, can at times be a particularly challenging test of a fire chief's management and leadership skills. It is, however, the essence of effective customer service and an integral component of fully understanding and responding to stakeholder expectations. Whereas your fire department will want to commit to meeting and, where possible, exceeding realistic stakeholder expectations, it will often be through the handling of complaints, concerns, and inquiries that you, as fire chief, will be afforded the opportunity to assist stakeholders in forming realistic expectations for the services that they expect and deserve to receive from your fire department. Take, for example, assisting new residents moving from a city with a fully paid fire department to your community where all fire department services are delivered through volunteer personnel in formulating reasonable expectations for your fire department. Obviously, realistic response times will vary under these two divergent staffing models. Previous relationships or associations that your fire department has established with its stakeholders through public education programs and community involvement will likewise have merit in this regard.

Issues to be Addressed

- Complaints
- Concerns
- Inquiries

Sources of Complaints, Concerns, and Inquiries

- Internal
- External

Your approach as fire chief, and that of your fire department, in handling complaints, concerns, and inquiries should be proactive and professional. It should be designed to support the effective and efficient receipt and resolution of all complaints, concerns, and inquiries in a proactive and positive manner. Upon receiving a complaint, identify the nature of the complaint and develop a full understanding of it. Attentive listening will enhance your understanding of the nature of and motivation behind the complaint. In receiving the complaint, apologize for any inconvenience that the person has experienced, and inquire as to how the individual lodging the complaint would like to see it resolved.

Characteristics of Effective Handling of Complaints, Concerns, and Inquiries

- Proactive
- Professional
- Responsive
- Timely

Essential Skills in Handling Complaints, Concerns, and Inquiries

- Interpersonal skills
- Listening skills
- Problem solving skills
- Empathy
- Commitment to customer service

In receiving and discussing the complaint with its originator, it is imperative that you not commit to any actions or make any promises until such time as you have had the opportunity to fully research the complaint and interview any and all involved and appropriate parties. All of your actions in investigating and handling the complaint should reflect due diligence and professionalism. It is always important to conduct yourself in a professional manner, and you should not tolerate abusive language or behavior on the part of the complainant.

It is important that you take each and every complaint seriously and consistently address them in a timely and professional manner. In most cases, request that the person presenting the complaint do so in writing. It is also important to realize that there will be times when you will receive inquiries or complaints that fall outside the scope of your responsibilities. In such cases, promptly refer the party to the appropriate individual or agency.

Relationship to Customer Service

Complaints, concerns, and inquiries provide valuable opportunities for your fire department to demonstrate its commitment to community and customer service.

The importance of your fire department providing the necessary level of service to meet and, where possible, exceed reasonable stakeholder expectations bears repeating, as does the fact that, whereas they can often represent a distraction that diverts your attention from other pressing issues, your diligent handling of complaints, concerns, and inquiries can be instrumental in positioning your fire department for present and future success, enhancing your success as fire chief. Your department's appropriate handling of these issues is directly related to a number of the expectations that stakeholders have for their fire department, as previously discussed in Lesson 10. These stakeholder expectations include: accessibility, courtesy, professionalism, responsiveness, and timeliness. It is essential to your success as fire chief and to that of your fire department that all personnel involved in receiving or addressing complaints, concerns, or inquiries, meet and, where possible, exceed the expectations of the department's stakeholders (figure 19–1).

Stakeholder Expectations Regarding Complaints, Concerns, and Inquiries

- Accessibility
- Courtesy
- Professionalism
- Responsiveness
- Timeliness

Figure 19–1. Fire officer discussing issue with fire department stakeholder

Job Aid 19 that appears at the end of this lesson is designed to assist you in analyzing how your fire department responds to complaints, concerns, and inquiries and in developing strategies, as appropriate, to improve your processes in this essential aspect of customer service.

Things to DO:

- Recognize the role that addressing complaints, concerns, and inquiries plays in determining your success as fire chief and that of your fire department, as well as in meeting and, where possible, exceeding realistic stakeholder expectations.

- Approach each complaint, concern or inquiry as a customer service opportunity.

- Use problem solving and interpersonal skills in addressing complaints and concerns.
- Be proactive and professional in addressing complaints, concerns, and inquiries.
- Respond to and resolve complaints, concerns, and inquiries in a timely manner.
- Use due diligence in the handling of complaints.
- Fully research all complaints before acting on them.

Things NOT to Do:

- Be reactive, rather than proactive, in addressing complaints, concerns, and inquiries.
- Procrastinate in responding to complaints, concerns, and inquiries.
- Fail to exhibit empathy when communicating with involved parties.
- Tolerate abusive language or behavior on the part of complaining parties.
- Make promises and commitments before fully researching a complaint.
- Handle complaints, concerns, and inquiries that do not fall within your responsibilities, rather than referring them to the appropriate entities.

Handling Complaints, Concerns, and Inquiries
(Job Aid 19)

1. Describe your fire department's policies, procedures, processes, and practices for receiving and addressing complaints, concerns, and inquiries.

2. Discuss your fire department's effectiveness and success in receiving and addressing complaints, concerns, and inquiries.

3. Discuss strategies that your fire department could use to enhance its effectiveness in handling complaints, concerns, and inquiries.

4. Discuss what you, as fire chief, could do to improve this important area of customer service within your fire department.

Lesson 20: Managing Projects

Role in Survival and Success

An all too frequent occurrence in many fire departments is when projects that are initiated, often with enthusiasm, fail to be properly implemented and then cannot achieve their intended outcomes. Some of these projects may represent worthwhile initiatives for the fire department to pursue, but many actually relate to mission-critical aspects of the fire department. Thus, the failure to effectively and efficiently initiate, manage, and complete these projects can undermine the fire department's success, as well as its ability to fully meet and, where possible, exceed the realistic expectations of its stakeholders.

Throughout your career as fire chief, you will encounter the initiation of many projects within the fire departments that you manage and lead. Some of these projects will culminate in the successful execution that leads to achieving the outcomes initially anticipated for the project, whereas others will represent significant failures for both you and your fire department. The management of each and every project should incorporate sound management principles and practice, particularly in terms of planning and execution, as well as stewardship in the use of your department's human, financial, and physical resources. The adoption and utilization of an effective process for selecting appropriate projects, assigning departmental resources to these projects, and supporting the successful implementation of selected initiatives will stand both you and your fire department in good stead.

What You Need to Know

Projects usually involve new or one-time initiatives that have an intended or desired outcome or result. Examples of new projects could include the creation of a public education division within the fire department; implementing a child car seat training and installation program; developing and implementing a new member orientation program; initiating a cadet membership program; and initiating a duty officer program. Representative one-time projects could include developing specifications for a new fire apparatus, ordering it, and preparing it to be placed into service; remodeling a fire station or building a new fire station; applying for a grant to purchase needed safety equipment; planning and conducting a regional fire officer training seminar; and providing the training and certification opportunities for all members of the fire department to attain a new level of professional certification.

It is important that the objectives of the projects that your fire department undertakes correspond with its established goals and mission, as well as with the expectations of its stakeholders. This should represent an essential consideration in the selection of projects for your fire department to pursue.

Projects usually have a specified timeframe over which they are to be completed. They may require the allocation and assignment of the department's human, financial, and physical resources. A budget, reflecting and justifying needed resources and their sources, should be developed in advance of selecting a project to pursue and should provide needed direction with respect to resource utilization throughout the life of the project.

Project

An organizational initiative with an intended outcome

Allocating Organizational Resources to Projects

- Financial resources
- Human resources
- Physical resources

An essential component to ensuring the success of a particular project is the selection of the departmental members and other individuals who will be assigned to the project. The decision regarding who should be charged with responsibility for the project is extremely important, as are the management and leadership skills of that individual. Successful project execution requires the use of effective delegation, along with the appropriate strategies to motivate and empower those individuals assigned to the project.

Project Management Process

Process utilized in managing a particular project

There are four components or stages in the successful management of a project. The first stage is the defining stage, during which the project is defined and structured. This stage is followed by the planning stage, during which plans are developed for the project. During the executing, or third, stage of project management, the major work activities of the project are undertaken and accomplished. It is during the final, or delivering stage, that the project is completed and implemented.

Stages in Project Management Process

1. Defining stage
2. Planning stage
3. Executing stage
4. Delivering stage

Defining Stage

Project management stage during which project is defined and structured

Planning Stage

Project management stage during which detailed plans are developed for the project

Executing Stage

Project management stage during which the major work on the project takes place

Delivering Stage

Project management stage during which the project is completed

The present and future success of your fire department will in large part be determined by your ability to select and support the appropriate projects for the members of your department to pursue. Ensure that the appropriate individuals are assigned to each and every project and are granted the necessary authority and resources to accomplish the responsibilities that they have been given with respect to the project (figure 20-1). Last, but by no means least, ensure that appropriate accountability measures are used so that the projects adopted by your fire department actually result in successful implementation. The importance of motivating and empowering the personnel assigned to each project should be obvious in determining the success of that project and its contribution to the overall success of your fire department.

Figure 20–1. Project planning meeting. (Courtesy of Latta White)

Job Aid 20 that appears at the end of this lesson is designed to be utilized in conceptualizing and planning a project within your fire department.

Things to DO:

- Recognize the role of project management in the successful management of your fire department.
- Adopt and utilize a project selection process to identify projects that correspond with and support the department's mission and goals and contribute to meeting and, where possible, exceeding realistic stakeholder expectations.
- Assign appropriate fire department personnel and other individuals to each project based on its specific needs.
- Utilize effective delegation in assigning project responsibilities and accompanying authority.

Things NOT to Do:

- Assign personnel who lack the requisite knowledge, skills, and interest to projects.
- Appoint the wrong person to manage and lead a project.
- Fail to determine and furnish the human, financial, and physical resources required to support a project.
- Fail to hold personnel accountable for project completion.
- Manage projects in an unorganized and unproductive manner.

Managing Fire Department Projects (Job Aid 20)

1. Discuss your fire department's effectiveness and success in managing projects that result in desired organizational outcomes.

2. Identify the reasons that project management initiatives have been successful or unsuccessful in your fire department.

3. Identify strategies that your fire department could use to enhance its success in the management of projects.

4. Discuss your role, as fire chief, in the successful management of projects within your fire department.

Lesson 21: Managing Change

Role in Survival and Success

One of the greatest challenges that you will continuously face as a fire chief involves the management of change. The environment in which your fire department exists and operates is dynamic and constantly changing. Change is thus inevitable, as, unfortunately, is resistance to change on the part of some members of your fire department. The successful enactment of your roles and responsibilities as fire chief will, therefore, dictate that you become a change agent, skilled in the management of change, including overcoming resistance to needed change.

An essential aspect of managing change is the creation of a positive organizational climate in which each and every member of the organization feels valued, appreciated, motivated, and empowered to fully contribute to the success of the fire department. Successful change management will require you, as fire chief, to recognize the need for change, along with the appropriate strategies to introduce and facilitate the adoption of change within your fire department, often in the presence of resistance on the part of fire department members and other stakeholders to accept and support the needed change.

Your success as a change agent will demand that you recognize that those who resist change are not bad people; rather they are just individuals with concerns or issues associated with the change that you have a responsibility to assist in addressing. In so doing, you will be preparing and positioning your fire department for success and survival and to better serve its stakeholders.

What You Need to Know

In recent years, numerous changes have taken place within the fire service, as well as within the larger environment in which fire departments operate. These changes have been and continue to be in crucial areas such as apparatus; equipment; technology; strategy and tactics; policies and procedures; and legislation, regulations, and standards. Whereas most fire chiefs have a fairly informed understanding of the changes in the fire service that have occurred to date, they recognize that in all likelihood the future holds even greater changes and challenges.

Examples of Change within the Fire Service

- Apparatus
- Equipment
- Legislation, regulations, and standards
- Policies and procedures
- Strategy and tactics
- Technology

Forces for change can originate from within your fire department, such as when a member suggests a new initiative, or from the outside, such as when standards or requirements change. Your fire department can approach change in either a proactive or a reactive manner. Reactive change results when a fire department does not plan for the future and is eventually forced by external developments to make certain changes. Proactive change results when the fire department takes ownership for its future and seeks to chart its course through strategic planning. Proactive, or planned, change is obviously preferable and in the best interest of your fire department and its present and future success.

Forces for Change

- Internal forces
- External forces

Organizational Approaches to Change

- Proactive
- Reactive

As with other aspects of successful fire department management, your fire department and its stakeholders will accrue significant benefits from the use of an organized change management process. Through such a process, and its logical sequence of activities, you will have a valuable tool at your disposal as you enact your role as a change agent. The change process that you utilize should incorporate the following sequential steps: recognize the need for change, establish goals for change, make a diagnosis, select a change technique or strategy, plan for implementation, implement the change, and evaluate and follow up as necessary.

Change Agent

The role of recognizing needed change and moving an organization and its members to implement the necessary change

Change Management Process

1. Recognize the need for change.
2. Establish goals for change.
3. Make a diagnosis.
4. Select a change technique.
5. Plan for implementation.
6. Implement the change.
7. Evaluate and follow-up as necessary.

Although this change management process will be instrumental to the effective management of change within your fire department, you will at times encounter departmental members and other stakeholders who demonstrate a resistance to accepting the proposed and needed change. As an astute fire chief, you must realize that it is imperative not to adopt a negative view of individuals who resist change, but to instead understand that they may simply be individuals with valid concerns that it is incumbent upon you to address through appropriate change strategies.

Members of your fire department and other organizational stakeholders may resist change for many reasons. The typical reasons why individuals resist change include uncertainty; surprise; inertia; misunderstanding; ignorance; lack of skills; lack of trust; fear of failure; emotional side effects; personality conflicts; self-interests; lack of tact; poor timing; different perceptions; threat to status or security; feelings of loss; or breakup of a workgroup.

Reasons Individuals Resist Change

- Breakup of workgroup
- Different perceptions
- Emotional side effects
- Fear of failure
- Feelings of loss
- Ignorance
- Inertia
- Lack of skills
- Lack of tact
- Lack of trust
- Misunderstanding
- Personality conflicts
- Poor timing
- Self-interests
- Threat to status or security
- Surprise
- Uncertainty

The six commonly recognized strategies for overcoming resistance to change are education and communication; participation and involvement; facilitation and support; negotiation and agreement; manipulation and co-option; and explicit and implicit coercion. Education and communication is appropriate when there is a lack of information or inaccurate information about the change, whereas participation and involvement is used in situations where change initiators lack the necessary information to formulate the change or others have significant power to resist it. Facilitation and support is an effective approach for problems associated with adjusting to the change. Negotiation and agreement is a realistic strategy when a powerful individual or group will lose as a result of the change. Manipulation and co-option should only be utilized when other strategies will not work, whereas explicit and implicit coercion should only be used when absolutely necessary as it can result in individuals resisting not only the change, but also resenting the change agent.

Change Management Strategies

- Education and communication
- Participation and involvement
- Facilitation and support
- Negotiation and agreement
- Manipulation and co-option
- Explicit and implicit coercion

Review of the stated reasons that individuals resist change, in light of the six available change strategies, should reveal that most types of resistance to change can be successfully addressed through the strategies of education and communication, participation and involvement, and facilitation and support. These three related strategies will serve you well throughout your career as a fire chief. Although there may be times when you consider utilizing more heavy-handed approaches, such as manipulation and co-option or even explicit and implicit coercion, consider these approaches only when absolutely necessary because their excessive use can significantly compromise your ability to manage and lead your fire department.

Your success as fire chief, as well as that of your fire department, will demand that you become a skilled and astute change agent capable of recognizing the need for change and mobilizing the support of fire department members and other stakeholders. In so doing, you will always want to value and respect those individuals who are resistant to a given change and to enable them to overcome their resistance through your skillful utilization of appropriate change strategies. A hallmark of your success in this essential dimension of managing and leading your fire department will be having the wisdom to implement change at a pace that is appropriate for your fire department and its members (figure 21–1).

Figure 21–1. Fire chief explaining procedural change to fire department members. (Courtesy of Broomall Fire Company)

Job Aid 21 that appears at the end of this lesson is designed to enable you to more successfully anticipate the need for change within your fire department and effectively and efficiently manage the implementation of change.

Things to DO:

- Recognize the challenges and importance of managing change within your fire department.
- Recognize that both change and resistance to change are inevitable.
- Understand the internal and external forces that lead to change within your fire department.
- Understand and successfully enact your role as a change agent within your fire department.
- Utilize a proactive approach to managing change within your fire department.

- Use appropriate strategies to overcome resistance to change.
- Recognize and address the need for change within your fire department.

Things NOT to Do:

- View individuals who resist change in a negative way.
- Utilize a reactive approach in managing change.
- Fail to approach change through an organized change management process.
- Fail to understand the reasons why individuals resist change and use appropriate strategies to overcome resistance.
- Routinely use manipulation and co-option, or explicit and implicit coercion in an attempt to overcome resistance to change.
- Implement change faster than appropriate in your fire department.

Managing Change in Your Fire Department (Job Aid 21)

1. Identify areas of change that present the greatest challenges to your fire department.

2. Discuss your fire department's success in the management of change.

3. Identify the reasons why members of your fire department resist change.

4. Discuss the strategies that your fire department should use to overcome resistance to change on the part of its members and other organizational stakeholders.

Lesson 22: Managing Conflict

Role in Survival and Success

Throughout your career as a fire chief, often at times when you least expect it, you will encounter the challenge of managing conflict. Conflict is defined as any situation in which incompatible goals, behaviors, emotions, or attitudes lead to disagreement between two or more parties. The presence of conflict can detract from your fire department's success in a number of ways, including compromising its ability to attain its mission, goals, and the expectations of its stakeholders.

The successful management of conflict will often test your management and leadership skills in significant ways. It is common that the same skills that are essential in managing change will also have value in the management of conflict. It is now recognized that under certain circumstances the existence of conflict can be a positive influence for needed change within your fire department. Your role in managing and leading your fire department will be to ensure that conflict is not allowed to rise to the level wherein it is counterproductive. Your attentiveness to conflict management and skills in this crucial area will play a major role in determining the success of your fire department, as well as your professional success as its fire chief.

What You Need to Know

The traditional view of conflict is that conflict is an undesirable thing that is always unhealthy. This mindset has been replaced by a contemporary view of conflict that, while recognizing that at times conflict will in fact represent an unhealthy influence within an organization, advocates that conflict can actually be healthy and serve a useful purpose when it results in the recognition of the need for change. Thus there will be times that conflict is unhealthy, but also times that it is healthy and serves a productive purpose.

Conflict

A dispute that exists between individuals, groups, or organizations

Traditional View of Conflict

Conflict is always unhealthy and undesirable.

Contemporary View of Conflict

At times, conflict can be healthy and serve a useful purpose.

The existence of excessive conflict, which is intense or continues for too long, will be an organizational problem that you will want to address in a timely manner. The absence of conflict, however, can also be problematic in that essential forces for change may not be present when this is the case. Your fire department can benefit from the existence of limited conflict in that conflict can be a positive and healthy force when it is of limited intensity and duration, but it is imperative that intense conflict, as well as conflict that is escalating or of long duration, be recognized as being unhealthy and be resolved in a timely manner.

Conflict Becomes Unhealthy When ...

- Conflict is intense.
- Conflict is escalating.
- Conflict is of long duration.
- Conflict is interfering with productive work.

As a fire chief, you may be required to deal with conflict that can occur at one or more of six possible levels. These levels are: intrapersonal, interpersonal, intragroup, intergroup, intraorganizational, and interorganizational. Intrapersonal conflict is conflict within an individual, whereas interpersonal conflict is conflict between individuals. An example of intrapersonal conflict would be a fire department member who is struggling with personal issues that are impacting his or her performance and involvement in the fire department. Interpersonal conflict would be present in a situation where two members of a fire department have a personality conflict that interferes with their working relationship.

Conflict within a group is called intragroup conflict, whereas conflict between groups is intergroup conflict. The presence of dissention within the fire officer ranks with respect to the roles and responsibilities of each officer is an example of intragroup conflict, whereas a rivalry that has turned into a conflict between two shifts or platoons within the fire department is an example of an intergroup conflict.

Intraorganizational conflict is conflict within an organization, and interorganizational conflict is conflict between organizations. Disputes, issues, or conflicts within the fire department are considered intraorganizational. When a conflict exists between two fire departments, for example with respect to competing for funding or response territory, an interorganizational conflict exists. A thorough reflection on your experiences in the fire service will likely reveal vivid and memorable examples of most of these types of conflicts.

Levels of Conflict

- Intrapersonal conflict
- Interpersonal conflict
- Intragroup conflict
- Intergroup conflict
- Intraorganizational conflict
- Interorganizational conflict

In managing conflict, you will benefit from the use of a three-phase approach to conflict management. The phases in this process are conflict prevention, conflict recognition, and conflict resolution. Conflict prevention attempts to prevent conflict through the use of appropriate strategies when and where possible, including providing a clear understanding of job responsibilities through written job descriptions in the interest of preventing role-related issues, such as role ambiguity and role conflict.

Conflict Management

1. Conflict prevention
2. Conflict recognition
3. Conflict resolution

The premise in conflict recognition is that unless conflict is identified and recognized, it will not be addressed. Thus, anticipating potential conflicts and being attentive to their occurrence is an essential ingredient of an effective conflict management program.

In the last step of the conflict management process, an attempt is made to address or resolve the conflict through the use of an appropriate conflict resolution strategy. Four strategies are available to assist you in the resolution of conflict within your fire department. Conflict avoidance seeks to ignore the conflict or to impose a solution. The intent of conflict diffusion is to smooth over the conflict. Conflict containment involves the use of bargaining or negotiation. Conflict confrontation, usually the most desirable among these conflict resolution strategies, seeks to proactively address and resolve the conflict through problem solving.

> **Conflict Resolution Strategies**
> - Conflict avoidance
> - Conflict diffusion
> - Conflict containment
> - Conflict confrontation

Your success as a fire chief and that of your fire department will in large part depend upon your understanding of the types and causes of conflicts and your knowledge and skills in effective conflict management. You will want to adopt a proactive approach to conflict management that recognizes the potential for conflict to be a healthy influence within your fire department, as well as its corresponding potential to compromise the success of your department.

You will find that good management and leadership practices will contribute to successful conflict management. Provide each member of your fire department with a clear understanding of his or her work assignment and responsibilities through a written job description. The use of cooperative, rather than competitive, strategies and rewards will likewise contribute to reducing conflict within your department. Obviously, the establishment and utilization of procedures for addressing conflict will be central to the successful management of conflict (figure 22–1).

Figure 22–1. Fire chief addressing conflict with fire officers. (Courtesy of Broomall Fire Company)

Your line of defense with respect to conflict should begin with the use of strategies designed to prevent conflict, followed by subsequent strategies designed to recognize conflict should it occur, and finally strategies designed to resolve conflict when appropriate. In managing conflict, it is important to never lose sight of the fact that, while often an unhealthy and counterproductive force, conflict that is limited in intensity and duration can serve a valuable role within your fire department in terms of calling attention to the need for change.

Job Aid 22 that appears at the end of this lesson is designed to assist you in evaluating your fire department's approach to managing conflict and identifying strategies that could enhance its effectiveness in this important area of responsibility.

Things to DO:

- Recognize the role of effective conflict management in your success as fire chief and that of your fire department.
- Recognize that conflict can be either a healthy or an unhealthy influence within your fire department.
- Do not allow conflict to escalate to a level where it is unhealthy and counterproductive.
- Recognize the types of conflict your fire department may experience.
- Seek to prevent conflict when and where possible.
- Be attentive in the interest of recognizing the occurrence of conflict.
- Resolve conflict in a timely manner utilizing an appropriate strategy.
- Adopt and practice a proactive and professional approach to conflict management.
- Establish and implement a procedure for managing conflict.

Things NOT to Do:

- View all conflict as unhealthy.
- Allow conflict to escalate to a level where it becomes unhealthy and counterproductive.
- Fail to address conflict that is intense, of long duration, or escalating in a timely manner.
- Procrastinate in managing conflict situations.
- Be reactive or unprofessional in managing conflict.

Managing Conflict in Your Fire Department (Job Aid 22)

1. Identify the areas of conflict that present the greatest challenges to your fire department.

2. Discuss your fire department's success in the management of conflict.

3. Discuss strategies that your fire department could use to manage conflict more effectively.

4. Discuss your role, as fire chief, in the management of conflict within your fire department.

Lesson 23: Managing and Leading With Integrity

Role in Survival and Success

The successful management and leadership of your fire department will demand that you ensure its compliance with all applicable laws, regulations, and standards at all times. Your effectiveness in this mission-critical aspect of your responsibilities as fire chief will serve as a standard by which the stakeholders of your fire department, including its members, will continually evaluate you. Full compliance in these areas will be essential to your success as fire chief, as well as that of the fire department that you have accepted the responsibility of managing and leading.

Although full compliance with relevant laws, regulations, and standards will provide a desired and essential foundation for your success and survival as fire chief, the expectations of your department's internal and external stakeholders will usually extend beyond simply obeying laws and complying with regulations. You will also be judged in terms of your ethics, integrity, and stewardship in enacting the roles and responsibilities of fire chief. A lack of attention to these essential aspects has the potential of undermining the success of your fire department and the respect and support it receives from the public and elected and appointed officials, as well as contributing to your reduced effectiveness as a fire chief, and in some cases ending what had previously been a promising career.

What You Need to Know

As the chief of your fire department, you have the ultimate, and often awesome, responsibility of managing and leading in a proactive manner that prepares the department for present and future success. Although there are many facets to the successful management and leadership of a contemporary fire department, an integral theme that threads through all of your decisions, actions, and activities should be the necessity of ensuring the complete compliance of your department at all times with relevant laws, regulations, and standards. Additionally, you will be expected to manage in a manner that consistently demonstrates ethical behavior, integrity, and stewardship. These attributes of your service as fire chief will stand your department in good stead in the present while properly preparing and positioning it for the future, and will serve as a tribute to your dedicated and professional service as fire chief for years to come.

Areas of Compliance

- Laws
- Regulations
- Standards

Behavioral Expectations

- Ethical behavior
- Honesty
- Integrity
- Stewardship

As fire chief, you will face numerous decisions each and every day, many of which will afford you the opportunity to manage with integrity or to fall short in this area of paramount importance to your department, as well as to you both personally and professionally. You will want to engage in ethical behavior at all times, wherein the decisions that you make and implement are made to the best of your ability, with due diligence, and do not result in harm to others.

It would be incorrect to claim that ethical decision making is always simple, straightforward, or easy, or that as fire chief you will not face situations that will require you to struggle with ethical issues. Whereas many of the decisions that you will need to make will have clear implications in terms of being ethical or unethical, you will discover ethical dilemmas can arise wherein a lack of clarity regarding the ethical appropriateness of a given situation will sometimes exist. These situations will require astute decision making, problem solving, and leadership skills on your part. You can better prepare yourself and the members of your fire department to successfully address ethical issues through the use of ethics training, the development and implementation of a code of conduct that incorporates ethical issues, and the use of ethics advisors (figures 23-1, 23-2, and 23-3).

Ethical Decision Making

Making decisions that are consistent with ethical principles and practices

Figure 23–1. Ethics training session. (Courtesy of Mike Matcho)

Code of Conduct

Written document that articulates an organization's expectations with respect to the conduct of its members

STANDARD OPERATING PROCEDURE

WILSON FIRE/RESCUE SERVICES

Subject Code of Conduct		Number A-4	Manual SOP	Effective Date 11/6/2003
Page 1 of 3	Prepared By SOP Committee	Revisions 2	Supersedes 7/18/2000	Approved Don Oliver Fire Chief
		Review Date 11/5/2003		

CODE OF CONDUCT

The Wilson Fire/Rescue Services is an integral part of a multi-faceted, pluralistic American community. As an organization, we want to ensure equity and inclusion in all our procedures and activities with each other and with all members of the public.

The following list of directives represents the conduct standards for members of the Wilson Fire/Rescue Services. The basis for these regulations is the following policy:

Every member of the organization is expected to operate in a highly self-disciplined manner and is responsible to regulate his/her own conduct in a positive, productive and mature way. Failure to do so will result in disciplinary action ranging from counseling to dismissal.

ALL MEMBERS SHALL:

1. Follow Operations Manuals and written directives of both the Wilson Fire/Rescue Services and the City of Wilson.

2. Use their training and capabilities to protect the public at all times, both on and off duty.

3. Work competently in their positions to cause all department programs to operate effectively.

4. Always conduct themselves to reflect credit on the organization.

5. Supervisors will manage in an effective, considerate manner: Subordinates will follow instructions in a positive, cooperative manner.

6. Always conduct themselves in a manner that creates good order within the organization.

7. Keep themselves informed to do their jobs effectively.

8. Be concerned and protective of each member's welfare.

9. Treat all, with respect and professional courtesy.

Figure 23–2. Fire department code of conduct. (Courtesy of Wilson Fire/Rescue Services)

Subject Code of Conduct	Number A-4	Effective Date 11/6/2003	Page 2 of 3

10. Be nice.

11. Operate safely and use good judgment.

12. Keep themselves physically fit.

13. Observe the work hours of their position.

14. Obey the law.

15. Be careful of department equipment and property.

MEMBERS SHALL NOT:

1. Engage in any activity that is detrimental to the organization.

2. Engage in a conflict of interest to the organization or use their positions with the organization for personal gain or influence.

3. Fight.

4. Abuse their sick leave.

5. Steal.

6. Use alcoholic beverages, debilitating drugs, or any substance which could impair their physical or mental capabilities while on duty.

7. Engage in any sexual activity while on duty.

8. In any way solicit free or discounted merchandise or services from any vendor.

   - Members may accept a professional courtesy discount that is normally offered to all other uniformed members in the community. (ie: non-alcoholic beverages, percentage discount on meals, etc.)

Figure 23–2. Fire department code of conduct. (Continued)

Figure 23–3. Fire department members meeting with ethics advisor. (Courtesy of Jerry Clark)

The fact that these decision-making situations can be problematic can be validated through a review of media coverage regarding fire and emergency services organizations and those entrusted with the responsibility for leading and managing them. All too frequently news stories, reported by traditional media sources such as newspapers, television, and radio, as well as the electronic media including websites, blogs, and social networking media, provide timely, comprehensive, and highly visible coverage regarding the conduct of fire and emergency services personnel with respect to honesty, ethics, integrity, and stewardship. These stories range from behavior that clearly violates laws and regulations, such as using the financial and other resources of the organization for one's own benefit, to those situations where decision makers simply made bad decisions without first considering their appropriateness and propriety. In all of these situations, both the reality of the involved situation and stakeholder perceptions regarding what took place have the potential to significantly compromise the reputation and standing of the fire department as well as all individuals involved. Whether or not you as fire chief were personally involved or knowledgeable of a wrongdoing, you will likely still be held responsible. The importance of proactively embracing issues associated with ensuring that your fire department is managed and led with integrity cannot therefore be overstated.

Your success and survival as a fire chief, as well as that of your fire department, require you to manage and lead with integrity at all times and in all the situations that you encounter. Integrity is the quality or state of sound moral principle and is a reflection on an individual's honesty, credibility, and sincerity. It is imperative that your conduct and behavior as fire chief reflect honesty and integrity at all times. A related aspect of integrity is understanding and avoiding situations that either represent or could be perceived as involving a conflict of interest. These pitfalls have snarled many fire chiefs and have quickly derailed their personal and professional standing, as well as their careers.

Integrity

Quality or state of being of sound moral principle

Conflict of Interest

Situation where there are tensions between an individual's responsibility to the organization and other conflicting demands or interests

A final aspect of managing with integrity that should serve to guide your decisions and actions is that of stewardship. The majority of the decisions you will make as a fire chief will have implications in terms of the allocation and use of your fire department's scarce resources. It is thus essential that you manage in a manner that makes the best utilization of your department's financial, human, and physical resources, thus demonstrating proper and appropriate stewardship. Resource decisions should always be guided by the established mission and goals of your department and the expectations of the stakeholders that it has the privilege of serving and protecting.

Stewardship

Personal responsibility for taking care of an organization's resources

Areas of Stewardship

- Financial resources
- Human resources
- Physical resources

The importance of adopting a proactive approach to the management and leadership of your fire department should be obvious, as should be the necessity of that approach in contributing to compliance with all relevant laws, regulations, and standards. Adopt such a stance early in your career and follow it throughout your tenure as a fire chief. The hallmark of your success will further demand the inclusion of a commitment to ethical behavior, integrity, and stewardship in all that you do as the fire chief. Such an enlightened approach will be a primary factor in determining your success as a fire chief and that of your fire department.

Job Aid 23 that appears at the end of this lesson is designed to assist you in developing a proactive stance and strategies with respect to managing and leading your fire department with integrity in a manner that ensures compliance with relevant laws, regulations, and standards, while demonstrating ethical behavior and proper stewardship for the organization's resources.

Things to DO:

- Recognize the importance of legal and regulatory compliance, ethics, integrity, and stewardship in positioning your fire department for present and future success.
- Recognize that the stakeholders of your fire department expect you to manage and lead with integrity.
- Ensure that your fire department complies with all relevant laws, regulations, and standards.
- Manage and lead with integrity at all times.
- Recognize that the stakeholders of your fire department expect you to exercise proper stewardship in the management and use of all departmental resources.
- Ensure that your decisions always lead to proper stewardship of departmental resources.
- Use your decision-making, problem-solving and leadership skills in addressing ethical decision-making dilemmas.
- Provide ethics training to members of your fire department.
- Establish and follow an appropriate code of ethical conduct.

Things NOT to Do:

- Compromise your integrity in your management approach.
- Fail to develop a complete understanding of the ethical implications of the decisions that you make.
- Assume that making ethical decisions is always an easy task.
- Fail to prepare and encourage department members to make ethical decisions.
- Make decisions that result in perceived or actual conflicts of interest.

Managing and Leading with Integrity (Job Aid 23)

1. Discuss your fire department's effectiveness with respect to conducting its operations in a manner that reflects ethical behavior and integrity.

2. Discuss the strategies that your fire department could use to ensure the integrity of its operations.

3. Describe what your fire department currently does to ensure that its members are enabled to enact their daily responsibilities and roles in an ethical manner.

4. Discuss your role, as fire chief, in ensuring that appropriate stewardship is exercised in all decisions regarding resource acquisition or use within your fire department.

Lesson 24: Managing Financial Resources

Role in Survival and Success

A challenge of growing significance that you will face as fire chief is that of managing the financial resources of your fire department. Fire departments, like all contemporary organizations, face many challenges related to the acquisition and use of financial and other resources. An ever-present reality is that these resources are scarce and must be effectively and efficiently managed. Although fire departments have faced such challenges for many years, the extent and implications of the budgetary and financial challenges facing a growing number of fire departments have the profound potential of compromising not only their present success, but also their future survival.

Your role as fire chief in the effective and efficient financial management of your department must be thoroughly understood and successfully enacted towards the end of ensuring the department's preparedness to continue to offer the services necessary to operate in an effective, efficient, and safe manner that meets and, where possible, exceeds the realistic expectations of its stakeholders. In enacting your role with respect to financial management, it is imperative that you utilize sound decision-making, problem-solving, and leadership skills that incorporate compliance with laws and regulations while demonstrating due diligence, ethical behavior, integrity, and stewardship. An underlying caveat to remember in managing the finances of your fire department is the importance of practicing fiscal responsibility and always recognizing that you are managing others people's money.

What You Need to Know

An integral responsibility that you will inherit as fire chief involves ensuring that your fire department is positioned to successfully acquire, and effectively and efficiently utilize, the financial resources required to enable it to engage in the effective, efficient, and safe delivery of services that the community it serves deserves and has come to expect. Use sound financial management practices that contribute to the effective and efficient acquisition and use of the funds required to secure and support necessary human and physical resources.

Your success as fire chief, as well as that of your fire department, will require that you develop a thorough understanding of the budget process utilized in your department or jurisdiction, including the parties involved in this process, as well as the necessary skills to effectively represent your department in budgeting processes and manage the budgets that are approved for your fire department. Budgets are designed to quantify and assign financial resources to organizational activities in accordance with the priorities of an organization. They serve as informational tools that enable an organization to "live within its financial means." In good times they are used to support new programs and initiatives; in challenging economic times they serve to restrict or control expenses. Budgets thus underpin and support all of the activities of your fire department.

> **Budget**
>
> A tool for managing an organization's resources

As a fire chief you will likely have responsibilities with respect to both operating and capital budgets. Operating budgets represent the revenues and expenses associated with a given year, whether expressed as a calendar year or a fiscal year. The expenses in an operating budget typically include such items as: wages and salaries, benefits, training, utilities, fuel, maintenance, materials, and supplies. Capital budgets are multi-year budgets designed to provide for major capital purchases and acquisitions that have associated high costs and an extended useful life. Capital projects include initiatives such as purchasing or refurbishing apparatus, constructing or remodeling a fire station, or making a major purchase of personal protective equipment. There may be special funding sources, including grants, loans, and bonds, available to support capital projects.

Types of Budgets

- Operating budget
- Capital budget

Operating Budget

A budget that reflects the operating expenses of an organization

Capital Budget

A budget that addresses the capital needs of an organization, such as land, buildings, apparatus, and equipment

Most operating budgets follow a line-item format, wherein each category of operating expense is broken down and identified separately and a designated allocation is made for each budget item. This is in contrast to a lump-sum budget, which simply provides a total allocation and leaves the details of developing a spending plan up to you as the fire chief. You will also develop and utilize program budgets to support designated programs, projects, or initiatives of your fire department.

Budget Formats

- Line-item budget
- Lump-sum budget
- Program budget

The process through which a budget is developed, approved, implemented, and administered is referred to as budgeting, with each of the steps in this process playing an important role in contributing to effective budgeting and financial management. The steps in the budget process are planning, preparation, presentation, approval, implementation, and administration and control. Budget planning involves gathering the necessary information to make and justify informed budget decisions through requesting input from appropriate departmental members with respect to the budgetary needs within their areas of responsibility.

Budgeting

The process through which budgets are developed, approved, implemented, and managed

Budget Process

1. Planning
2. Preparation
3. Presentation
4. Approval
5. Implementation
6. Administration and control

Budget preparation begins with the identification and evaluation of revenue sources. In considering revenues, you will often find it prudent to research and pursue alternate revenue sources to supplement the traditional funding sources of your fire department. You can enhance your awareness of potential new revenue sources through review of *Funding Alternatives for Fire and Emergency Services*, a publication of the United States Fire Administration that is available to your fire department. The second task in budget preparation is to identify anticipated expenses, or expenditures, along with their accompanying justifications. Your professionalism, integrity, and stewardship will be exemplified through the development of realistic and justifiable budget requests that are based on a comprehensive understanding of the economic realities and financial situation of your fire department and the jurisdiction that it serves.

Revenues

Financial resources received by an organization

Expenses (Expenditures)

Costs associated with the operation of an organization

After you have developed a set of reasonable revenue and expense projections, you face the challenge of developing a balanced budget that reconciles projected revenues and anticipated expenses. This part of budget preparation may prove to require difficult deliberations, particularly in challenging economic times, but it is essential to effective budgeting and sound financial management. In developing a budget it is better to underestimate revenues and overestimate expenses. For each item in the budget, whether an anticipated revenue or a projected expense, provide an accompanying justification (figure 24-1).

Figure 24–1. Fire department budget meeting. (Courtesy of Bob Sullivan)

The preparation of a budget is logically followed by a budget presentation, during which the proposed budget is presented to interested stakeholders in accordance with the budgeting requirements and practices of your department or jurisdiction. The budget approval process that follows will vary among organizations and jurisdictions and may involve an internal and/or external budget review. In some cases you will be required to participate in public budget hearings on behalf of your fire department (figure 24–2). Upon its approval, you will receive the necessary authorization to utilize approved funding in support of your fire department and its operations during the period covered by the budget.

Figure 24–2. Fire chief presenting budget request at public meeting. (Courtesy of Chapel Hill Fire Department)

The final step in the budgeting process, budget administration and control, is designed to ensure that your fire department operates in accordance with its approved budget. Various budget control tools will provide invaluable information and insights as you seek to administer your department's budget properly and professionally, thus demonstrating the desired and expected stewardship with respect to the management of the financial resources of your fire department.

Many fire chiefs lack a comfort level or interest in the financial management of their fire department, but relinquishing your responsibility for and active involvement in this mission-critical area is never advisable. Through your active involvement in the budgeting processes and financial management of your fire department, you enhance your department's ability to meet, and often exceed, the realistic expectations of its stakeholders, as well as preparing and positioning it for present and future success. Your use of a proactive, yet prudent, approach will enhance your success in budgeting and financial management in the interest of ensuring that your fire department has the necessary resources to accomplish its mission and meet and, where possible, exceed the reasonable expectations of its stakeholders.

Job Aids 24A and 24B that appear at the end of this lesson are designed to assist you in formulating an understanding of the budget process within your fire department in the interest of enhancing your effectiveness in ensuring that your department has the necessary resources to fulfill its mission and meet and, where possible, exceed stakeholder expectations.

Things to DO:

- Recognize the role of sound financial management in preparing and positioning your fire department for present and future success, as well as in determining your effectiveness as its fire chief.

- Develop a thorough understanding of the budgeting process utilized in your department and jurisdiction, as well as the players in this process.

- Seek to fully understand your role in budgeting and other financial management activities related to your fire department.

- Assume an active role in your department's budgeting and financial management.

- Ensure that your fire department is able to acquire necessary funding.

- Understand the differences between operating and capital budgets.

- Prepare program budgets to secure needed funding for new projects or programs.

- Make informed financial management decisions based on accurate and relevant information.

- Recognize the reality of scarce resources and seek to manage all resources in an effective and efficient manner.

- Seek the involvement of appropriate members of your fire department in the determination of resource needs to support their areas of responsibility.

- Seek to accurately identify and evaluate projected revenues and anticipated expenses, recognizing that in preparing a budget it is better to underestimate revenues and overestimate expenses than to do the reverse.

- Investigate and seek new or alternate revenue sources.

- Justify all budget requests.

- Always demonstrate ethical behavior, integrity, stewardship, and due diligence in the financial management of your fire department.

Things NOT to Do:

- Assume a limited role in your department's budgeting and financial management.

- Fail to pursue and secure necessary funding for your fire department.

- Make financial management decisions based on incomplete or inaccurate information.

- Overestimate projected revenues or underestimate anticipated expenses in order to balance a budget.

- Prepare and submit budget requests that are unrealistic based on the current economic or financial situation.

- Fail to attend and participate in budget presentations or hearings.

- Fail to properly justify all budget requests.

- Fail to utilize budgetary controls during budget administration.

- Lose sight of the fact that you are making decisions with respect to other people's money.

Fire Department Budgeting (Job Aid 24A)

1. Describe the budget processes of your fire department and the municipality or municipalities that it serves, including the parties to these processes.

2. Discuss the traditional role of the fire chief in budgeting and other financial management activities within your fire department.

3. Discuss how your fire department could enhance its effectiveness in budgeting and financial management.

4. Discuss, as appropriate, any changes you could make in your role as fire chief in budgeting and financial management that would be beneficial to your fire department and the community that it serves.

Developing a Program Budget
(Job Aid 24B)

1. Discuss projects for which your fire department has developed a program budget and the associated outcomes.

2. Identify a proposed project for your fire department that would require a program budget to ensure the necessary acquisition and utilization of funds to successfully complete the project and realize its intended outcomes.

3. Identify the necessary funds required for this project, as well as potential funding sources.

4. Identify the major expense categories associated with this project.

Lesson 25: Marketing Your Fire Department

Role in Survival and Success

The work that you do as fire chief on a daily basis, along with that of the other members of your fire department, is designed to prepare and position your fire department to operate in an effective, efficient, and safe manner that meets and, where possible, exceeds the realistic expectations of its stakeholders. Most fire departments recognize this responsibility and do all within their power to prepare for and meet the needs of the jurisdiction that they serve and protect, and consequently the expectations of stakeholders.

As successful as many fire departments are in enacting the aforementioned responsibilities, most fail to fully recognize the need to promote or "market" their department to its various stakeholder groups. This is unfortunate from several perspectives. The first is that not pursuing an engaged interaction or relationship with fire department stakeholders through a proactive marketing strategy results in a lost opportunity on the part of your fire department to "sell itself," as well as to "tell its story" to those it serves. An accompanying consequence of not developing and utilizing an aggressive marketing strategy is that it likewise represents a lost opportunity wherein the fire department could "make its case" to the public and other stakeholders in terms of the support it requires to effectively, efficiently, and safely deliver the services necessary to meet and, where possible, exceed their expectations.

What You Need to Know

As a progressive fire chief interested in enhancing your department's preparedness and success, pursue the appropriate opportunities to communicate with your department's stakeholders. Your department will reap significant, and often unimaginable, benefits through the development and implementation of a proactive marketing plan. Your communication and leadership skills will prove instrumental to your success and that of your department in terms of marketing itself. Although it is possible to view such a marketing approach as somewhat self-serving on the part of your fire department, it is really anything but that in that the intent of a properly developed and implemented marketing program should be to enhance your department's ability to better serve its stakeholders.

Marketing

Activities undertaken by an organization to communicate with and inform stakeholders about the organization and the products and/or services that it provides

Marketing Plan

A written document that delineates an organization's marketing strategies and activities

The use of marketing will allow your organization to enable its stakeholders to fully understand and appreciate the services that your fire department provides, as well as the challenges that it faces. Your fire department, therefore, has much to gain through the use of a marketing program that includes strategies and messages specifically targeted to its various target audiences.

Benefits of Marketing Your Fire Department

- "Sell your fire department"
- "Tell your fire department's story"
- "Make your fire department's case"

In addition to the many traditional promotional approaches available to your fire department, including letters, brochures, and posters, your department will benefit from marketing strategies that incorporate direct personal contact and interaction with its stakeholders, such as open houses and community presentations. New technologies, such as websites and various forms of social media, will also prove invaluable in extending the marketing reach and effectiveness of your fire department to certain stakeholder groups.

Traditional Promotional Approaches

- Brochures
- Letters
- Posters
- Community presentations
- Open houses

Contemporary Marketing Approaches

- Websites
- Blogs
- Various forms of social networking media

It would be remiss not to acknowledge your fire department's public education role and the impact that such programs and initiatives can and should have on community risk reduction. Public education programs should be viewed as an essential component of your fire department's service and a major potential determinant of community risk reduction. These programs and initiatives can be additionally instrumental in supporting your fire department's marketing program and you will always want to take a few minutes at the start of any presentation to department stakeholders, regardless of whether a public education presentation to a community group or the presentation of a budget request to elected officials, to provide a brief overview of your fire department, the jurisdiction that it serves, and the services that it provides (figure 25–1).

Public Education Programs

Provide a valuable opportunity to market your fire department to program participants.

Figure 25–1. Fire department members marketing fire department during public event. (Courtesy of Philadelphia Fire Department)

Establishing a marketing program in your fire department may seem like a rather significant and daunting task, but you will find that it really is not. There are many resources available through governmental agencies, including the United States Fire Administration (USFA), and professional organizations to assist you in this endeavor. Develop working relationships with your local media, and through the use of press releases, secure their assistance in helping your department to "share its good news" and communicate with its various stakeholder groups.

Job Aid 25 that appears at the end of this lesson is designed to assist you in considering the effectiveness of your fire department's present marketing initiatives and in formulating, as appropriate, new strategies to more successfully market your fire department to its stakeholders.

Things to DO:

- Recognize the important role that marketing can play in positioning your fire department for present and future success.
- Recognize the possible benefits to your stakeholders of the fire department marketing itself.
- Use marketing to "sell your fire department," "tell its story," and "make its case" for support.
- Adopt a proactive and aggressive approach to marketing your fire department.
- Develop and implement a marketing program tailored to the needs of various target audiences.
- Select and utilize appropriate marketing tools in your marketing activities.
- Take the opportunity to market and promote your fire department during all presentations to the public, elected officials, and other stakeholder groups.
- Utilize new technologies, such as the Internet and social media, to reach selected stakeholder audiences.
- Develop and nurture relationships with local media.
- Issue press releases to enlist media assistance in marketing your fire department.

Things NOT to Do:

- Miss opportunities as a result of your fire department not properly marketing itself.
- Adopt a reactive approach to marketing your fire department.
- Fail to investigate and use all available assistance in support of your marketing efforts.

Marketing Your Fire Department
(Job Aid 25)

1. Describe your fire department's marketing efforts and experience to date.

2. Describe the benefits that your fire department could realize through an effective marketing program.

3. Identify the target audiences that you would address in developing a marketing program for your fire department.

4. Identify the marketing approaches and strategies your fire department could utilize to reach the identified target audiences.

Lesson 26: Working with Stakeholder Groups

Role in Survival and Success

Throughout your career as a fire chief, you will find yourself in situations that necessitate that you work with the stakeholders of your fire department. These interactions will range from formal to informal. Some will involve a single individual; others will involve a group of departmental stakeholders. In each of these engagements, it is imperative that you conduct yourself with the utmost of professionalism and integrity while demonstrating empathy with respect to the issues, interests, concerns, and expectations of each stakeholder group.

Developing a cooperative working relationship with the various stakeholders of your fire department will stand you in good stead as fire chief and will also position your fire department to realize present and future success. As stakeholders, each involved individual or group has a vested interest in your success as fire chief and that of your fire department and thus can be instrumental in providing the necessary support and assistance to enable you and your fire department to succeed in meeting, and ideally exceeding, their expectations.

What You Need to Know

In working with departmental stakeholders, you will realize the importance of sound leadership, communication, decision-making, and problem-solving skills. Many of the management skills discussed throughout this book will prove essential as you work with the various stakeholders of your fire department. Professionalism and integrity will be key ingredients of your success whether working with a single stakeholder or a group of your stakeholders. Although there may be times that you find yourself in spontaneous discussion or interaction with departmental stakeholders, you will always want to be prepared for these engagements through maintaining a current understanding and appreciation of the expectations and issues of each stakeholder group. Although there are general guidelines that you will want to follow when dealing with your department's stakeholders, you will be well served by understanding certain specific aspects of working with the various stakeholder groups of your fire department.

Stakeholder Groups

- Fire department members
- The public
- Elected and appointed officials
- The media
- Other fire and emergency services agencies and organizations

The members of your fire department are considered internal stakeholders and comprise a primary stakeholder group that you will never want to fail to consider as you manage and lead the department. Ensure that you are always professional, honest, and credible in your dealings with the members of your department. The importance of integrity and dedication in your management/leadership approach cannot be overstated in that such an approach will yield the necessary relationship of trust and mutual respect wherein department members will afford you the opportunity to lead, as opposed to simply manage, using the authority that accompanies your position as fire chief.

The work of your fire department will be performed through each and every member enacting his or her specified roles and responsibilities in an orchestrated manner. When working with the members of your fire department, ensure that your actions and activities contribute to a supportive organizational climate that leads to motivation, empowerment, job satisfaction, and job performance, and thus to effective personnel recruitment and retention (figure 26–1).

As the fire chief, ensure that each and every member of your fire department fully understands his or her role and the important contribution that his or her work makes to the overall success of their fire department. Likewise, provide all of the resources that the members of your department need to effectively, efficiently, and safely enact their roles and responsibilities. One of your most important responsibilities to the members of your fire department is to afford each and every individual the opportunity to fully develop and utilize their talents.

Figure 26–1. Fire chief interacting with personnel at fire station.

A second primary stakeholder group with which you can expect to have frequent, ongoing interaction is the public. It is important to always remember the degree to which the general public depends on the essential services provided by your fire department and thus the numerous expectations that they have for your department. These expectations were discussed in Lesson 10. You will find it extremely beneficial to develop cordial and cooperative working relationships with individual members of the public, as well as with organized community groups (figure 26–2).

An integral aspect of working with the public may involve educating these stakeholders so as to enable them to develop realistic expectations for your fire department. Understanding their expectations and being committed to meeting and, where possible, exceeding realistic stakeholder expectations will enhance the public support that your fire department receives and consequentially contribute to its present and future success and survival.

Figure 26–2. Fire chief meeting with community group. (Courtesy of Wilson Fire/Rescue Services)

Whereas elected and appointed officials in most cases will represent some of the most pleasant and pleasurable stakeholders with whom you will have the pleasure to work, that will not always be the case. Ensure that you always keep your elected and appointed officials informed so that they never find themselves in a situation where they are unaware or uninformed (figure 26-3). Be honest with these stakeholders at all times, even with respect to the response capabilities and readiness of your fire department. It is essential that you involve elected and appointed officials and assist them in assuming appropriate "ownership" for the public safety of your community. If there are resources that you need in order to effectively, efficiently, and safely serve your community, it is imperative that you inform the elected and appointed officials of your jurisdiction of these resource needs, but only ask for the resources that your fire department really needs and can justify. Note the similarities with respect to your desire that the stakeholders of your fire department likewise develop realistic and reasonable expectations for your fire department.

Figure 26–3. Fire chief meeting with municipal manager. (Courtesy of Latta White)

It is almost assured that you will have many opportunities to work or interact with the media throughout your service as fire chief. Many of these situations will involve the media covering a breaking news story in which your department has become involved. In such situations, recognize that the news media's job is to always get a story and that they will do so through official and/or unofficial sources, as necessary. Take the opportunity to ensure that the news coverage with respect to incidents to which your department responds, as well as events in which the fire department is involved, is reported accurately. As the fire chief, you will benefit from appointing and utilizing the services of a public information officer, as well as making yourself available to speak with the media when appropriate (figure 26–4).

Figure 26–4. Fire commissioner meeting with reporters. (Courtesy of Philadelphia Fire Department)

The positive working relationships that you develop with local reporters and their news organizations can be instrumental in affording your fire department the opportunity to inform and persuade the public and other relevant stakeholder groups on issues that are important to the fire department and influence its ability to accomplish its goals and achieve its mission. An important caveat in working with the media is to recognize the importance of referring the media to appropriate sources in those cases where you either cannot or should not address a particular matter.

Although it is easy at times to view the media as being against the fire department, this is usually not the case at all. Through a proactive approach to working with the media that incorporates developing appropriate working relationships, making yourself available at appropriate times to address issues that are within your expertise, honoring commitments that you make to the media, and finally being a true professional in all that you do, both you and your department will realize the success that can come through working with the media.

A final stakeholder group deserving of mention is the other fire and emergency services organizations or agencies that depend upon your organization. Whether your fire department is receiving assistance from or assisting these organizations, you will quickly discover that you are all in this together (figure 26–5). An essential dimension of working with other organizations and agencies is the importance of fully understanding, respecting, and managing in accordance with their established roles, responsibilities, and authorities. You should never put another organization in a problematic position in terms of compromising or undermining its ability to fulfill its designated roles and responsibilities. The hallmarks of your work with these organizations should be professionalism, honesty, and integrity.

Figure 26–5. Fire chief meeting with emergency medical services chief

The challenges and issues associated with working with the various stakeholder groups of your fire department represent an appropriate final issue to examine in this book, given the importance of meeting and, where possible, exceeding the realistic expectations of your fire department's stakeholders. The management survival skills discussed throughout this book will prepare and position you as fire chief to successfully represent your fire department in the numerous and frequent dealings that you will have with your department's stakeholders. As with other issues discussed earlier, a proactive, professional approach that is grounded in ethical behavior, integrity, mutual respect, and understanding will be instrumental in preparing your fire department to succeed in the present and future, and to meet and, where possible, exceed the realistic expectations of each group of stakeholders.

Job Aid 26 that appears at the end of this lesson is designed to assist you in identifying relevant stakeholder groups and in formulating strategies for enhancing your working relationships with these groups.

Things to DO:

- Recognize the importance of working with stakeholder groups as you enact your responsibilities as fire chief.
- Recognize that various stakeholder groups can influence the present and future success of your fire department.
- Emphasize professionalism and integrity in all of your dealings with the stakeholders of your fire department.
- Seek to motivate and empower all members of your fire department.
- Manage and lead in a manner that creates a positive and supportive organizational climate.
- Ensure that all personnel know their roles and responsibilities and are provided the necessary resources to enact them.
- Ensure that all members of your department are given the opportunity to fully develop and use their talents.
- Seek to develop trust and mutual respect with departmental stakeholders.
- Recognize the degree to which the public depends on your fire department.

- Develop cooperative relationships with individual members of the public and organized community groups.
- Purpose to fully understand and meet and, where possible, exceed the reasonable expectations of the public.
- Recognize the potential for challenges and issues when working with elected and appointed officials.
- Be honest in all dealings with elected and appointed officials.
- Bring resource deficiencies to the attention of elected and appointed officials.
- Justify and be reasonable when making requests for additional resources.
- Encourage elected officials to take ownership for the public safety of their community.
- Recognize that the media's job is to "get the story."
- Understand that the media will use both official and unofficial sources in covering a story.
- Take the opportunity to promote your fire department through the media and to ensure that news coverage is accurate and fair.
- Recognize the limits of your expertise and, when appropriate, refer the media to more appropriate sources.
- Understand the roles, responsibilities, and authorities of other fire and emergency service organizations and agencies and never put them in a compromising or problematic position.
- Recognize that your dealings with stakeholders are a reflection on you, your fire department, and its members.

Things NOT to Do:

- Fail to understand the issues and expectations of interest to each group of stakeholders.
- Fail to view members of your fire department as an important stakeholder group.
- Fail to practice honesty, ethics, and integrity in all dealings with the members of your department.
- Fail to keep elected and appointed officials informed on important issues.
- Fail to develop positive working relationships with the media.
- Comment on issues that are beyond your expertise or on which you should not comment.
- Engage in interactions with stakeholders that diminish your professional reputation or that of your fire department.

Working with Stakeholder Groups (Job Aid 26)

1. Discuss your fire department's present effectiveness, as well as strategies that it could use to enhance its effectiveness, in working with fire department members.

2. Discuss your fire department's present effectiveness, as well as strategies that it could use to enhance its effectiveness, in working with the public.

3. Discuss your fire department's present effectiveness, as well as strategies that it could use to enhance its effectiveness, in working with elected and appointed officials.

4. Discuss your fire department's present effectiveness, as well as strategies that it could use to enhance its effectiveness, in working with the media.

Final Thoughts

This completes our pilgrimage. Hopefully, you have gained insights that will enhance your knowledge, skills, and attitudes with respect to successfully enacting your present and future roles and responsibilities as a contemporary fire chief. Serving in this mission-critical role within your fire department and community is a privilege that very few individuals are afforded throughout their fire service careers. The management survival skills shared in this book were intended to contribute to your success as a fire chief and that of the fire department(s) that you will have the opportunity to manage and lead.

It is important to remember that it all comes down to people. Your fire department's internal and external stakeholders are individuals who, based on their situation and perspective, will have a set of expectations for you as fire chief and for your fire department. It is important for you to develop positive working relationships with each stakeholder group, seeking to understand their expectations and, when necessary, assist them in revisiting unrealistic expectations. It is your charge as fire chief to meet and, where possible, exceed the reasonable expectations of the various stakeholder groups for you and for your fire department.

Always remember that your professional success and that of your department will derive from the commitment and dedication of the members of your fire department. They deserve honesty, integrity, and credibility in all of your dealings with them. Likewise, you have a responsibility in terms of stewardship for the fire department, which at some future time you will hand over to the person who follows you as fire chief.

Among your most important responsibilities is ensuring that you have afforded all of your personnel the opportunity to fully develop and utilize their talents. In so doing, you are preparing the cadre of individuals who, as fire officers, will continue to lead your fire department with the same dedication, commitment, and passion as you have. When you think about it, there could be no greater tribute to your service as fire chief.

Index

A

accessibility, for fire and emergency services, 94

accountability
through delegation, 68–69, 72, 112
by fire chief, 21
through job description, 130
for project, 182, 184

aging, of population, 12, 77

Alderfer, Clayton, 111

American Council on Education (ACE), 52, 56

apparatus, firefighting
assignment of, 33
challenges for, 15
changes in, 187–188
maintenance of, 1, 130
readiness of, 25, 130
as resource, 34
self-contained breathing apparatus for, 130
strengths and weaknesses of, 80
SWOT analysis for, 79, 81, 86, 88

appraisal. *See* evaluation

assignments, by fire chief, 33, 70, 72

Assistance to Firefighter's Grant Program, 130

authority
through delegation, 68–69, 72–73, 112
of fire chief, 41, 73
by management, 43, 47
over personnel, 41
for projects, 182, 183

B

budget
acquisition of funds for, 213–214
approval of, 218
controlling of, 219
economy and, 77
effectiveness of, 222
ethics for, 221
for fire department, 6, 16, 25–26, 29, 77, 213–222
format for, 215
funding sources for, 130, 214, 220–221
justification of, 221
knowledge and, 214
management of, 213–215, 220
member involvement in, 221
operating and capital funds for, 214–215, 220
presentation of, 218–219
priorities for, 214
process for, 216–218, 220, 222
resources for, 16, 26, 29, 77, 213–215
rewards and incentives in, 108

C

career
as chief officer, 1, 3–5
evaluation throughout, 129–130, 132
staffing models as, 99–100, 129
transition in, 1, 4–9, 22, 32, 132

Census Bureau, U.S., 12–13, 15

certification
Department of Homeland Security for, 55
for fire officer, 50, 53–55, 57–59, 123, 168
goals for, 60, 84
International Fire Service Accreditation Congress for, 55
National Board on Fire Service Professional Qualifications for, 55
for professional development, 5, 7, 50, 53
Standard for Fire Officer Professional Qualifications for, 54
status and self-esteem through, 110–111

fire chief in, 18, 43, 102, 139, 142, 203, 210–211
personnel in, 102, 203
recruitment and retention in, 102, 203
with regulation, 18, 102, 139, 142, 203–211
conflict
of interest, 209, 211
leadership and, 15, 25, 26, 124–125, 127–128, 167, 195–202
levels of, 197–198
management of, 15, 25–26, 124–125, 127–128, 167, 195–202
as negative force, 195–197, 200–201
prevention of, 198, 200, 201
proactive approach to, 199, 201
as productive, 195–196, 200–201
recognition of, 198, 200–201
resolution of, 198–199, 200–201
in roles, 124–125, 127–128
consistency
in communication, 155
in discipline, 142
in evaluation, 135
by fire department, 94
consolidation, of fire departments, 15, 77, 80
controlling
for budget, 219
by fire chief, 30, 35–37, 39, 42, 157
for incident, 35–36
types of, 35–36
convenience, of fire and emergency services, 94
courtesy, for fire and emergency services, 94
credentials. *See* certification
culture, social environment of
as external environment, 77–78
of fire department, 77–78, 80, 107, 110, 113, 114, 149, 232, 238
as threat, 80
customer service, by fire department, 171–175, 177

D

decision making
challenges in, 157–158
classical model for, 159–160, 165–166
commitment to, 161
cost-effectiveness of, 159, 161
effectiveness in, 157, 162, 164
through empowerment, 112
ethics of, 205, 211
evaluation of, 159–160, 163
by fire chief, 15, 22–26, 30–31, 37, 42, 157–158, 161–164, 204, 231
goals in, 160
as informed, 157–159, 162
leadership for, 15, 22–26, 30–31, 37, 42, 157–163, 164, 204, 231
maximizing or optimizing in, 160, 162
monitoring of, 163
necessity for, 163
policy for, 148, 155
priorities in, 162
risk in, 157–158
roles for, 15, 22–26, 30–31, 37, 42, 157–158, 161–164, 204, 231
satisficing in, 160–161, 162
strategies for, 157, 159, 161–164
timeliness of, 161
delegation
accountability through, 68–69, 72, 112
as appropriate, 71, 73
as assignment, 72
authority through, 68–69, 72–73, 112
effectiveness in, 74
empowerment through, 112–113
by fire chief, 5–6, 21, 26, 27, 29, 37, 38, 67–71, 74
by job description, 69
to personnel, 71–72
as professional development, 72
for project, 180–182, 183–184
responsibility through, 68–69, 72, 112
strategies for, 74
time management through, 67, 72–73

recruitment and retention and,
104, 105, 108
of stakeholders, 7, 11–12, 14, 17,
21, 25, 26, 32, 41, 47, 49, 64,
75–77, 86, 91–95, 99, 114,
117, 139, 171–172, 174–175,
204, 219–220, 225, 231, 240,
243
of volunteers, 100

F

facilities
as resource, 34
SWOT analysis for, 79, 81, 86, 88
Fire and Emergency Services Higher
Education (FESHE), 55–56
fire chief
accountability by, 21
appointment of, 7
assignments by, 33, 70, 72
authority of, 41, 73
behavioral expectations for, 204
budget by, 6, 16, 25–26, 29, 77,
213–222
career transition to, 1, 4–9, 22,
32, 132
challenges for, 11, 20, 49, 65, 95
as change agent, 16, 18, 85, 86,
187–193
commitment by, 29, 243
communication by, 15, 24, 30–31,
37, 42, 125, 145–156, 225–
226, 231
community needs and, 21, 23, 25,
227, 239
conflict management by, 15,
25–26, 124–125, 127–128,
167, 195–202
controlling by, 30, 35–37, 39, 42,
157
decision making by, 15, 22–26,
30–31, 37, 42, 157–158, 161–
164, 204, 231
delegation by, 5–6, 21, 26, 27, 29,
37, 38, 67–71, 74
directing by, 30, 34, 37, 39, 42, 157
discipline by, 139–144
empowerment by, 15, 18, 37, 72,
99, 107, 112–113, 114

as entrepreneur, 25, 26
environmental trends and, 77
ethics of, 7–8, 17–18, 29, 38,
46–47, 134, 203–209, 231,
238, 240
evaluation by, 129–131, 137
evaluation of, 21, 26, 203
expectations for, 7, 11–12, 14, 17,
21, 25–26, 32, 41, 47, 49, 64,
75–77, 86, 91–95, 99, 114,
117, 139, 171–172, 174–175,
204, 219–220, 225, 231, 240,
243
financial and physical resources
and, 11, 12, 15, 16, 25, 26, 29,
34, 213–215
future direction by, 75–76, 78, 86
at incident scene, 3, 21, 35, 42
informational role for, 3, 22–24,
26, 151–153, 235
initiative by, 15, 31, 32
inside and outside roles by, 22,
27, 125
job descriptions by, 15, 21, 69,
105, 117, 119–124
judgment of, 29
leadership by, 5, 6, 11, 13, 15–16,
18, 22, 23, 26, 27, 41, 44–45,
47, 48, 49, 75, 114, 157, 203,
204, 231
in legal compliance, 18, 43, 102,
139, 142, 203, 210–211
as liaison, 23, 26
management skills of, 15–16,
25–26, 29–30, 37, 41, 44–45,
48–49, 61–64, 67, 72–73,
75, 124–125, 127–128, 167,
179–185, 195–203
media interaction with, 21,
235–236
as mentor, 7
motivation by, 15, 18, 37, 43, 72,
99, 107
negotiation by, 25, 26
organizing by, 30, 37, 39, 42, 75,
157
personnel and, 15, 16, 233
politics and, 15, 77–78, 80
powers as, 47–48
practical experience of, 6
preparation for, 49